Recent Topics in Mass Spectrometry

Recent Topics in Mass Spectrometry

Edited by

ROWLAND IVOR REED

University of Glasgow, Scotland

GORDON AND BREACH SCIENCE PUBLISHERS

New York   London   Paris

Foreword

The recent developments in mass spectrometry are reflected in the articles appearing in this book which arise from a NATO Study Institute of Mass Spectrometry held in Lisbon, August 1969.

This course, as those held previously, was primarily concerned with the more elementary aspects of mass spectrometric theory and practice. However, many of the lectures hinted at more advanced concepts and it is these aspects which are reported here. The papers are contributed by acknowledged experts in the field and may thus be considered as authoritative. They should prove of great value to workers in the field.

An operation of this kind depends both upon the good will and the willing collaboration of a great many people. High upon this list must stand the instrument manufacturers and it is with pleasure that I acknowledge the assistance of A.E.I. Scientific Apparatus Ltd., Manchester, England; Finnigan Corporation Ltd., Palo Alto, California; Bendix Corporation Ltd., Rochester, New York; Field-Tech Ltd., London, England, Vacuum Products Ltd. and Via-Vac Ltd., Gatwick, England.

As always, the success of such a venture must depend upon the lecturers, the associated staff, and the publishers. All have responded to make the operation highly successful; the work of the lecturers and publishers can be seen in this work, but without the able assistance of the other staff this project would have been stillborn.

It is also with great pleasure that I acknowledge the enormous assistance freely offered by the members of the University of Lisbon at all levels; that of the University of Glasgow of whose help I stood in need and the staffs of Galleon-W.T.A. Ltd. and T.A.P. for travel arrangements.

Finally I must express my great gratitude to NATO for the provision of the bulk of the funds needed to mount this operation, to the Calouste Gulbenkian Foundation, and the Instituto de Alto Cultura for further financial assistance; to the representatives of all these institutions who visited us, to the Portugese Travel Association for their help and finally to the President of the Institute, Professor K. Jacobson who in spite of the crushing burden of his normal commitments threw himself whole-heartedly into this project and did much to ensure its success.

ROWLAND IVOR REED

Glasgow   *Director*

Contents

NATO Institute Lecturers

Hemel Hempstead, Herts., England

G. BALL, Finnigan Instruments Ltd.

J. E. COLLIN, Institut de Chimie, Université de Liège, Belgium.

R. W. CRAIG, Vacuum Generators Ltd., East Grinstead, Sussex, England.

N. R. DALY, United Kingdom Atomic Energy Authority, Aldermaston, Berkshire, England.

S. EVANS, A.E.I. Scientific Apparatus Ltd., Urmston, Manchester, England.

M. A. A. FERREIRA (Mrs.), The Calouste Gulbenkian Laboratory for Mass-Spectrometry, Lisbon, Portugal.

M. F. LARANJEIRA, The Calouste Gulbenkian Laboratory for Mass-Spectrometry, Lisbon, Portugal.

G. R. LESTER, Imperial Chemical Industries Ltd., Blackley, Manchester, England.

K. LEVSEN, Institut für Physikalische Chemie, Bonn, W. Germany.

A. J. LUCTHE, Bendix Scientific Instruments Ltd., Rochester, N.Y., U.S.A.

F. MENDES, The University of Lisbon, Portugal.

C. A. MERRIT Jr., Pioneering Research Laboratory, Natick, Mass., U.S.A.

A. QUAYLE, Shell Research, Thornton Research Centre, Chester, England.

M. T. ROBERT-LOPES (Miss), The University of Lisbon, Portugal.

A. G. SHARKEY JR., Bureau of Mines, Pittsburgh, Pennsylvania, U.S.A.

*H. J. SVEC, U.S.A.E.C., Iowa State University, Ames, Iowa, U.S.A.

M. E. WACKS, The University of Arizona, Tucson, Arizona, U.S.A.

---

* Unable to come. He has provided a manuscript for this book.

Multiply charged ions

MORTON E. WACKS and WARD M. SCOTT

The University of Arizona, Tucson, Arizona 85721, USA

INTRODUCTION

Interest in the occurrence of multiply charged molecular ions and rare gas ions that could be expected in the atmosphere of planets and stars prompted a survey of experimental data available on such ions by Mohler in 1964.[1] Rare gas ions of multiple charge have been identified and classified spectroscopically for all the rare gases.[2] Research on multiple ionization of atoms and molecules by electron collision was stimulated by theoretical studies by Wannier[3] and Geltman.[4]

Other information on multiply charged ions in mass spectrometry was sparse, and the most detailed discussion of multiple ionization prior to Mohler's review was given by Dorman and Morrison[5] (1961). Interest in multiply charged ions in organic compounds has been evident since the report of doubly charged polyatomic ions by Conrad in 1930. Recently the application of doubly charged spectra to structural problems in organic mass spectrometry was recognized.[6,7,8] Some examples of doubly charged ions in 70 volt mass spectra whose intensities are greater than the corresponding singly charged species may be found in the literature.[6,7,10,11,12] Abundant doubly charged ions have been observed in the mass spectra of several substituted diphenylmethanes,[6] phenylenediamenes,[7] fused ring aromatics,[13] and triphenylmethanes.[14]

Instances of the decomposition of one doubly charged ion to another doubly charged ion and a neutral species, i.e. $AB^{++} \rightarrow A^{++} + B^0$, have

1

been observed[15] (supported by metastable ion transitions). Recently meta-stable ion transitions involving the decomposition of triply charged ions have been reported.[16] The doubly charged fragmentation sequences in these types of compounds provide additional information to that of the singly charged dissociative-ionization processes which can be of use in structural problems[8] once the processes stabilizing multiply charged ions are under-stood.

DISCUSSION

Studies of processes of n-fold ionization in the monatomic gases induced by electron impact have shown that the ionization probabilities vary above threshold as the nth power of the excess electron energy.[17-20] The range over which this dependence is found to apply varies with n but is in no case less than 5 eV. The probability for multiple ionization of molecules should similarly be dependent upon the nth power of excess electron energy; how-ever, the process will be complicated by the presence of vibrational levels within each electronic state, which will modify the shape of the ionization efficiency curve for the ion in question.

Transitions induced by electron impact involving the removal of two or more electrons from a molecule are likely to cause severe rearrangement in electronic structure, much more so than in the removal of one electron. Direct electrostatic interactions will contribute much more to the description of the chemical binding in the multiply charged ions than in singly charged ions and it is not unexpected that the dissociative pathways for the multiply charged species may differ significantly from those of the corresponding singly charged species.

The formation of multiply charged organic ions may result from direction ionization, i.e. $AB + e^- \rightarrow AB^{++} + 3e^-$, by autoionization from an excited level of a singly charged species (charge separation process) i.e. $AB^+ \rightarrow AB^{++} + e^-$ or ion pair formation from the singly charged species i.e. $ABC^+ \rightarrow AB^{++} + C^-$. Recently metastable transitions for such processes were found in 1,3,5-trichlorobenzene, while charge separation processes were found to occur in 6,6-diphenyl-2,4-pentadienylidene-1.[25]

Stabilization of ionic species having more than one electron removed in the ionization or dissociative-ionization process requires reducing to a mini-mum the coulombic repulsion between the charges. Coulombic forces may not then provide sufficient additional energy to that contained in the ion as a result of the ionization process and the thermal energy present in its various degrees of freedom to cause dissociation into two or more singly charged ionic species. The abundance of multiply charged ions has been shown to be increased by the presence of Π-electron systems and non-bonding electrons on heteroatoms.[9,14,21-23] The concept of charge localization directing fragmentation[24] has recently been applied successfully to explain the contrasting behavior of singly and doubly ionized ions of 4-methoxycarbonylbenzene-1,2-dicarboxylic anhydride.[8] The fragmentation scheme for the singly charged dissociative pathway is,

$$C_9H_3O_4^+ \xrightarrow{-CO} C_8H_3O_3^+$$

$$\text{(} -\text{OMe)} \nearrow \qquad \downarrow -CO_2$$

$$C_{10}H_6O_5^+ \qquad C_7H_3O^+ \xrightarrow{-CO} C_6H_3^+$$

$$-CO_2 \searrow \qquad \downarrow -OMe$$

$$C_9H_6O_3^+ \xrightarrow{-CO} C_8H_6O_2^+$$

While that for the doubly charged species is

$$C_{10}H_6O_5^{++} \xrightarrow{-CO_2} C_9H_6O_3^{++} \xrightarrow{-OME} C_8H_3O_2^{++}$$

$$\swarrow -CO$$

$$C_6H_3^{++} \xrightarrow{-CO} C_7H_3O^{++}$$

In the singly charged dissociative-ionization pathway there are competing primary reactions involving the anhydride and methoxycarbonyl groups. Either primary dissociation is followed by loss of CO completing the expected two step sequence characteristic of the functional group prior to the first step of the sequence associated with the second group. In the doubly charged fragmentation sequence the first step is loss of CO_2 followed by decomposition dominated by the methoxycarbonyl moiety, the anhydride group not competing effectively in this sequence. This is expected in the doubly charged decomposition if charge localization in the functional groups is postulated in the doubly charged molecular ion.

Multiply charged ions may be identified in low resolution mass spectra[14] of compounds of C, H, and O on the basis of the intensities of their naturally occurring isotopic peaks which occur at non-integral mass-to-charge ratios. Compounds containing an odd number of nitrogen atoms may yield doubly charged ions at non-integral mass-to-charge ratios. Unambiguous identification of multiply charged species can be accomplished by employing high resolution mass spectrometric techniques. This has recently been demonstrated by Vouros and Biemann in the mass spectra of some substituted phenylenediamines.[7]

Phenylenediamines

In the phenylenediamines the principle stabilization of multiply charged onic species is attributed to charge localization at the heteroatom sites. This ocalization is evident from the examination of ionization energies of benzene, isopropyl benzene, t-butyl benzene, aniline and phenol. The presence of the amino group lowers the ionization energy[26] of benzene by 1.5 eV indicating that loss of a non-bonding electron from the heteroatom leads to ion formation. Separation of charge by localization at each nitrogen atom in compounds of the type R_1R_2N—C_6H_4—NR_3R_4 would therefore account for the observed occurrence of doubly charged ions in these compounds. However, localization of charge by the heteroatoms in related phenolic structures or by hyperconjugation in the t-butyl or isopropyl groups would not be as pronounced (these groups lower the ionization energy of benzene by approximately 0.5 eV). Therefore the reduced intensities of the doubly charged species in the mass spectra of the oxygen analogs of the phenylenediamines i.e. the low abundances of doubly charged ions in the ortho, meta, and para diethoxy benzenes,[7] are not unexpected.

Diphenylmethanes

Abundant doubly charged ions were observed in the mass spectra of som t-butyl substituted dihydroxy diphenylmethanes.[14] Metastable ion transition were found supporting the decomposition pathway:

$$(M-30)^{++} \xrightarrow{-28} (M-58)^{++} \xrightarrow{-28} (M-86)^{++}$$

in 3,3′,5,5′-tetra-t-butyl-4,4′-dihydroxy diphenylmethane, and 3,3′-di-t-butyl-5,5′-dimethyl-4,4′-dihydroxy diphenylmethane.[6] The $(M-30)^{++}$ ion would appear to be first formed as in an analogous manner to that shown in the decomposition of t-butyl benzene by Rylander and Meyerson.[27] Stabilization of the two positive charges in this ion is achieved by delocalization of each charge by each aromatic nucleus. Interaction of the charges is prevented by the non-conjugated methylene bridge, and charge separation is maintained at a distance sufficient so that the coulombic repulsion energy is reduced below that necessary to cause dissociation into two singly charged ions.

Table 1 presents the percent of the total ionization due to doubly charged ions for some diphenyl methanes. The doubly charged ions in these compounds are of much greater intensity than their singly charged counterparts. In 3,3'-di-t-butyl-5,5'-dimethyl-4,4'-dihydroxy diphenylmethane, these ratios are: $I(M-30)^{++}/I(M-30)^{+} > 10$; $I(M-58)^{++}/I(M-58)^{+} > 20$; and $I(M-86)^{++}/I(M-86)^{+} > 70$.

Table 1   Percentage of total ionization due to doubly charged ions in the mass spectra of some diphenyl methanes

Compound	$\Sigma_{++}/\Sigma I_{39}$ (%)
A. 3,3',5,5'-tetra-t-butyl-4,4'-dihydroxy diphenylmethane	11.0
B. 3,5-di-t-butyl-3',5'-dimethoxy-4,4'-dihydroxy diphenylmethane	5.2
C. 3,3'-di-t-butyl-5,5'-dimethyl-4,4'-dihydroxy diphenylmethane	13.1
D. 3,5-di-t-butyl-3',5'-dimethyl-4,4'-dihydroxy diphenylmethane	5.8

Formation of the doubly charged species can be visualized in terms of the ionization-dissociation processes observed in t-butyl benzene, i.e. loss of a methyl radical may occur from each of two t-butyl groups if they are on different phenyl groups of the ion, thus yielding an $(M-30)^{++}$ ion. Whereas if both t-butyl groups are located on the same aromatic nucleus (as in compounds B and D) formation of an $(M-30)^{++}$ ion would lead to both charges on the same aromatic nucleus and result in a marked decrease in the stability of the doubly charged ion due to disruptive coulombic repulsion forces. Therefore, in compounds B and D, the $(M-30)^{++}$ ions are extremely small while intense $(M-15)^{++}$ and $(M-43)^{++}$ ions are easily observed. Thus the doubly charged spectra of these compounds provide useful information in structure elucidation, in addition to that found in the singly charged spectra.

This stabilization of charge by formation at the t-butyl site, followed by migration to the aromatic nucleus across the pseudo-conjugated system in each aromatic moiety prompted a search for a compound which might provide intense triply charged ions. The compound 3,3′,3″,5,5′,5″-hexa-t-butyl-4,4′,4″-trihydroxy triphenylmethane (E) exhibited this phenomena as expected.

(E)

Table 2   Principal multiply charged ions in the 70 volt mass spectrum of 3,3′,3″,5,5′,5″-hexa-t-butyl-4,4′,4″-trihydroxy triphenylmethane

Ion	$(M-30)^{++}$	$(M-58)^{++}$	$(M-45)^{+++}$	$(M-73)^{+++}$	$(M-101)^{+++}$	$(M-129)^{+++}$
R.A.	25.6	9.2	2.1	1.3	0.9	2.5

The relative abundances of the principal doubly and triply charged ions in the 70 volt mass spectrum of this compound are listed in Table 2.

Triphenylmethanes

Examination of the mass spectra of organic dyes being used in radiation dosimetry[28] experiments showed that multiply charged ions were formed easily in malachite green substituted at the central carbon atom. Table 3 lists the major singly charged ions (greater than 20% of the base peak) and the doubly charged ions observed in substituted malachite green dyes (F) where X = H, OH, OCH_3, and OC_2H_5, and crystal violet methoxide (G).

The principal mechanism for stabilization of these doubly charged species appears to be similar to that observed in the diphenylmethanes for the most part, i.e. separation of charge by localization at the nitrogen atoms on two

(F)

(G)

aromatic moieties separated by a non-conjugated methylene bridge. This can be seen by comparison of the total intensities of the multiply charged species in malachite green methoxide (13% of the base peak) and crystal violet methoxide (37% of the base peak). The addition of a third nitrogen atom on the remaining aromatic ring provides three ways in which two charges can be accomodated in the crystal violet methoxide, compared to one in the malachite green methoxide, in agreement with the observed ratio of the doubly charged species (essentially 3:1). The same effect can be seen in the total ionization due to doubly charged species in the hexa-t-butyl substituted

Table 3 Relative abundance of major ions of some triphenylmethane dyes (70 volt spectra)

Compound	Singly Charged	Doubly Charged
Malachite Green–X		
X = H	$M^+ = 100$; $(M-77)^+ = 64$	$M^{++} = 18$; $(M-78)^{++} = 12$
X = OH	$M^+ = 100$; $(M-17)^+ = 82$ $(M-77)^+ = 97$; $m/e(148)^+ = 45$ $m/e (105)^+ = 22$	$(M-18)^{++} = 9$ $(M-94)^{++} = 10$
X = OCH$_3$	$M^+ = 32$; $(M-31)^+ = 100$	$(M-31)^{++} = 7$ $(M-32)^{++} = 6$
X = OC$_2$H$_5$	$M^+ = 20$; $(M-45)^+ = 100$	$(M-45)^{++} = 12$ $(M-46)^{++} = 12$
Crystal violet methoxide	$M^+ = 17$; $(M-31)^+ = 100$	$(M-32)^{++} = 20$ $(M-31)^{++} = 17$

triphenylmethane when compared to the tetra-t-butyl substituted diphenyl-methane.

A second mechanism for stabilization of doubly charged ions seems to be present in these compounds. There is a tendency to form doubly charged species corresponding to the loss of both non-nitrogen containing substituents of the central carbon atom. This can result in a stable, highly conjugated structure of an allenic nature. This diquinoid ion structure could have a positive charge on each nitrogen atom with the planes of the two quinoid groups mutually perpendicular, a requirement necessary to maintain maximum separation of the charges.

SUMMARY AND CONCLUSION

Stabilization of multiple charges on ionic species produced by 70 volt electrons in a mass spectrometer occurs by the mechanisms of charge separation and charge dispersion. Charge separation may be brought about by localization of each charge at specific heteroatomic sites in the ion (e.g. in the phenylenediamines and triphenylmethane dyes) or by delocalization of each charge on separate non-interacting aromatic systems as observed in the t-butyl substituted diphenylmethanes. In addition, it is possible to stabilize ions containing more than one electronic charge even in completely conjugated systems such as the fused-ring aromatics which exhibit multiply charged spectra. Charges in these systems tend to separate by coulombic repulsion in such a manner that the effective centers of charge are as far apart as possible.

The driving forces which dominate fragmentation pathways in singly charged species also direct the dissociation pathways of multiply charged species, e.g. formation of even electron ions, increased conjugation and aromaticity (providing charge separation is still possible), etc. Such factors readily result in doubly charged ions which are significantly more intense than their singly charged counterparts. The resulting dissociation pathways observed for multiply charged ions provide a further internal mass spectrum from which structural information can be deduced.

References

1. F.L. Mohler, "Survey of Multiply Charged Ions", *NBS Tech. Note* **243**, U.S. Government Printing Office, 0-732-135, Washington, D.C. (1964).
2. C.E. Moore, *Atomic Energy Levels III*, Table 34, NBS Circular 467 (1958).

3. G.Wannier, *Phys. Rev.* **90**, 817 (1953).
4. S.Geltman, *Phys. Rev.* **102**, 171 (1956).
5. F.H.Dorman and J.D.Morrison, *J. Chem. Phys.* **35**, 575 (1961).
6. J.D.Fitzpatrick, W.M.Scott, C.Steelink, and M.E.Wacks, *Int. J. Mass Spectr. Ion Phys.* **1**, 415 (1968).
7. P.Vouros and K.Biemann, *Org. Mass Spectrometry*, **2**, 375 (1969).
8. S.Meyerson, I.Puskas, and E.K.Fields, *Chem. Comm.*, 346 (1969).
9. R.Conrad, *J. Physik*, **31**, 888 (1930).
10. S.Meyerson, *App. Spectroscopy*, **9**, 120 (1955).
11. J.H.Beynon, *Mass Spectrometry and its Application to Organic Chemistry*, Elsevier Press, Amsterdam, Holland (1960).
12. J.A.McCloskey, R.N.Stillwell, and A.M.Lawson, *Analyt. Chem.* **40**, 233 (1968); G.H.Draffan, R.N.Stillwell, and J.A.McCloskey, *Org. Mass Spectrometry*, **1**, 669 (1968).
13. M.E.Wacks and V.H.Dibeler, *J. Chem. Phys.*, **31**, 1557 (1959); M.E.Wacks, *J.Chem. Phys.*, **41**, 1661 (1964).
14. M.E.Wacks and W.M.Scott, *International Conference on Mass Spectroscopy*, Paper III C-2, Kyoto, Japan (September 8–12, 1969).
15. S.Meyerson and R.W.Vander Haar, *J. Chem. Phys.*, **37**, 2458 (1962).
16. K.R.Jennings and A.F.Whiting, *Chem. Comm.*, 820 (1967).
17. V.H.Dibeler and R.M.Reese, *J. Chem. Phys.*, **31**, 283 (1959).
18. J.D.Morrison and A.J.C.Nicholson, *J. Chem. Phys.*, **31**, 1320 (1959).
19. M.Krauss, R.M.Reese, and V.H.Dibeler, *J. Research NBS*, **63A**, 201 (1959).
20. F.H.Dorman, J.D.Morrison, and A.J.C.Nicholson, *J. Chem. Phys.*, **31**, 1335 (1959).
21. F.L.Mohler, V.H.Dibeler, and R.M.Reese, *J. Chem. Phys.*, **22**, 394 (1954).
22. E.J.Gallegos, *J. Phys. Chem.*, **71**, 1647 (1967).
23. F.H.Field and J.L.Franklin, *Electron Inpact Phenomena*, Academic Press, New York (1957).
24. F.W.McLafferty, *Chem. Comm.*, 78 (1966).
25. J.Seibl, *Org. Mass Spectrometry*, **2**, 1033 (1969).
26. J.L.Franklin, J.G.Dillard, H.M.Rosenstock, J.T.Herron, K.Draxl, and F.H.Field, "Ionization Potentials and Heat of Formation of Gaseous Positive Ions", *NSRDR-NBS* **26**, U.S. Government Printing Office, Washington, D.C. (1969).
27. P.N.Rylander and S.Meyerson, *J. Am. Chem. Soc.*, **78**, 5799 (1956).
28. M.Wankerl, M.E.Wacks, and L.Harrah, *Trans. Am. Nuclear Soc.*, **12**, 60 (1969).

Energetics of ionization and dissociation processes by electron impact

M. A. ALMOSTER FERREIRA

Laboratorio Calouste Gulbenkian de Espectrometria de Massa e Fisica Molecular, Lisboa

1 INTRODUCTION

The development of new experimental techniques, and the increased accuracy of theoretical calculations due to the use of computers, are making the knowledge of the energies related to a molecule either by itself or when involved in a chemical reaction, become more and more precise. For this development, the mass spectrometric methods are taking a very important position, especially through measurements of ionization potentials and appearance potentials. In many cases they provide the only available data on the energies of molecules, radicals or ions, especially for organic compounds, through the measurements that can be made for the various ions observed in their mass spectra.

When a diatomic molecule AB in the gas phase is bombarded by electrons of a certain energy, ionization may occur, either with or without dissociation leading to the formation of positive or negative ions. The formation of positive ions is more abundant by a factor of about 10^3 with respect to the negative ions[1], and because of it, positive ions may be considered more important. Positive ions will be considered first and negative ions will be treated separately.

The formation of positive ions by electron impact can be represented by:

$$AB + e \underset{\text{(fast)}}{\rightarrow} AB^+ + 2e$$

$$AB + e \underset{\text{(fast)}}{\rightarrow} A^+ + B + 2e$$

11

For polyatomic molecules several different fragments can be obtained either ionized or neutral.

When the energetics of ionization and dissociation processes are to be considered, some details about the knowledge of ionization and appearance potentials of atoms, molecules and radicals are of capital importance.

The first ionization potential of an atom or molecule is defined as the energy required to remove a valence electron from the lowest occupied atomic or molecular orbital of the neutral particle to infinite distance, giving rise to the correspondent atomic or molecular ion in the ground state. For a molecule it corresponds to the energy difference between the ground vibrational levels of the lowest electronic states of the molecule and the molecular ion, and is called the adiabatic ionization potential.

When ionization by electron impact takes place, an electron with sufficient energy will bombard the molecule and will be allowed to interact with it for a period of about 10^{-14} to 10^{-15} seconds[2]. The time for ionization is then too short when compared with the time required for the quickest vibration—which is larger than 10^{-3} sec. As a consequence the conditions for ionization are such that they can be described by the Franck–Condon principle.

The conditions before and after the ionization will be represented by points of the potential energy curves which represent the molecule in the ground state before the transition and the molecular ion also in the ground state after the transition, which lie on a parallel to the energy axis. The ionization potential measured in this way is referred to as the vertical ionization potential.

The behaviour of some systems obeying the Franck–Condon principle is represented in Figure 1. It should be noted that it can be expected that the vertical ionization potential does not always correspond to the adiabatic ionization potential. The vertical ionization potential represents an upper limit of adiabatic or true values, and it can be concluded that the appearance potential of a molecular ion will usually be greater than the ionization potential of the molecule, although with most molecules the difference is smaller than 0.1 eV.

The appearance potential of an ion is related to the minimum electron energy necessary to promote the decomposition of the molecule, so that the special ion will be formed.

When a diatomic molecule is ionized the appearance potential of the ion obtained can then be considered to be the heat of reaction that gives rise to its formation. But if a diatomic molecule is ionized and at the same time

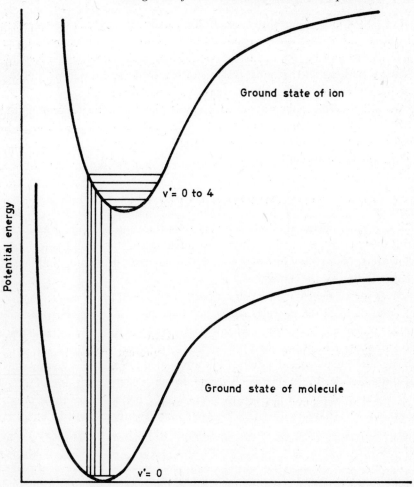

Ground state of ion

v'= 0 to 4

Potential energy

Ground state of molecule

v'= 0

Internuclear distance

Figure 1   Vertical transitions in ionization[12]

excited, the molecular ion will decompose into two fragments carrying excess energy. In this case the Franck–Condon region is entirely above the decomposition asymptote, and the measured appearance potential will be equal to the heat of reaction plus the sum of the translational energies of the products. This translational energy will be equal to the difference between the excitation energy of the molecule and the heat of reaction, and is usually referred to as the excess energy[3].

For polyatomic systems, part or all of the excess energy can be necessary for the vibration modes of the products, and a simple relation as for diatomic molecules cannot be found, since potential energy surfaces are involved[4].

One would expect that most fragment ions would be formed with excess energy, but experience has shown that this is not the case.

We will not refer to the ways in which ionization and appearance potential measurements are done since they have been already mentioned[5]. We will see how they can be used to obtain information concerning energetics of the ionization and dissociation processes.

2   CALCULATION OF BOND ENERGIES

The energy of chemical bonds is a very important quantity for a molecule since most chemical reactions involve the breaking of some bonds and, at the same time, the making of new ones. It is in fact one of the most important characteristics of the molecule.

Mass spectrometric methods have been largely used in the determination of bond dissociation energies since they have proved to be of more wide applicability than, for instance, kinetic methods even if the latter can generally provide more reliable results.

The bonds between individual atoms are related to the lowest energy state of the molecule. If the energy of the molecule is increased some of these bonds can be broken. The energy necessary to break one of these bonds leading to the dissociation of the molecule into two parts, will be equal to the difference between the heats of formation of the fission products and the heat of formation of the undissociated molecule.

If a molecule contains two or more identical bonds we have to distinguish between the energy necessary to break one of these bonds and the mean bond energy. For instance, for methane the bond energy $D(H_3C—H)$ is about 102 kcal and the mean bond energy $D(C—H)$ is about 99 kcal[6]. This is due to modifications in atomic configurations occurring when the molecule dissociates, which affect the energies of the remaining atoms. This problem can be very well understood by means of the mass spectrometric method since measurements can be made for a molecule corresponding to the successive removal of the several atoms of the molecule, and the appearance potential of each of the fragments obtained can be determined.

If several molecules which by dissociation give rise to the same fragment ion are considered and if in each case the appearance potential of the fragment ion is determined, different values are obtained. This implies that appearance potential measurements can be used to calculate bond dissociation energies. For which purpose one is interested in the process

$$AB + e \rightarrow A^+ + B + 2e \tag{I}$$

If the energy of the process is E and the ionization potential of the radical A is known, the following relation holds:

$$D(A - B) = E - IP(A) \tag{II}$$

Between the energy E and the appearance potential of the ion A^+ the following relationship holds:

$$AP(A^+) = E + E_1 + E_2 + K_1 + K_2 \tag{III}$$

where E_1 and K_1 are the excitation energy and the kinetic energy of the radical B, E_2 and K_2 the corresponding quantities for A^+.

We assume that the minimum value of the appearance potential corresponds to the zero excitation energies of A^+ and B; on the other hand, since the kinetic energy will be distributed between A^+ and B in accordance with the laws of conservation of energy and angular momentum[7], another form of equation (III) is:

$$AP(A^+) = E + \frac{M_1 + M_2}{M_1} K_2 \tag{IV}$$

where M_1 and M_2 are the molecular masses of fragments A and B.

Determining the kinetic energy of the ion A^+ (through a retarding potential method[8], for instance) E can be calculated by equation (IV), and the required bond energy $D(A - B)$ will be given by equation (II).

In practice we assume that both excitation and translation energies of the fragments are zero. This is usually true for excitation energies, and kinetic energies are very often lower than 0.2 eV so that, when any information on excitation and kinetic energies is missing, these two quantities are simply ignored. Nevertheless one must write

$$D(A - B) \leqslant AP(A^+) - IP(A)$$

to imply that, in principle, what is really calculated is an upper limit of the bond energy.

The theory of mass spectra shows that rates of reactions which may occur during fragmentation, are dependent on the actual activation energy of the reaction, which may be equal to the endothermicity of the process plus an

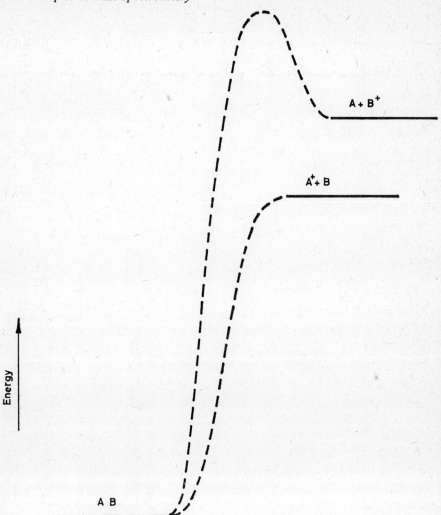

Figure 2   A simple illustration of Stevenson's rule

additional energy barrier as shown in Figure 2. As a consequence of studies of that kind, a principle known as Stevenson's rule[9] has been established which refers to the possible fragmentation processes of a molecular ion. According to this rule when dissociative ionization occurs the charge usually resides on the fragment having the lowest ionization potential. That means that if $IP(A)$ is smaller than $IP(B)$, the decomposition $A^+ + B$ is more probable, and if the reaction producing $B^+ + A$ occurs, the products will in some way carry excess energy. From here an important conclusion can be drawn: when determining bond energies only the reaction leading to the formation of the ion which, as a radical, has a lower ionization potential may be used.

As an illustration, let us consider the calculation of bond dissociation energy of the SC—S bond in CS_2 [10]. Two possibilities can be considered

$$CS_2 + e \rightarrow CS + S^+ + 2e$$

$$CS_2 + e \rightarrow CS^+ + S + 2e$$

The first of these two possibilities gives

$$D\,(CS\text{—}S) \leqslant AP\,(S^+) - IP\,(S) \leqslant 4.4 \text{ eV mole}^{-1}$$

The second one

$$D\,(CS\text{—}S) \leqslant AP\,(CS^+) - IP\,(CS) \leqslant 1.7 \text{ eV mole}^{-1}$$

But since

$$IP\,(S) < IP\,(CS)$$

it is the first, the value 4.4 eV mole^{-1} that must be taken as the bond dissociation energy $D\,(CS\text{—}S)$.

Sometimes it is possible to measure directly by electron impact the ionization potential of the neutral fragment which enters into

$$D\,(A - B) = E - IP\,(A)$$

The radicals have to be submitted as individual particles to ionization, and one way of obtaining the radicals is by pyrolisis of halogen substituted molecules which easily loose the halogen. The pyrolisis has to be carried out close to the ionization chamber so that the radicals may enter it practically without collisions[11].

When this procedure is not possible ionization potential of radicals can be calculated by means of

$$AP\,(B^+) = D\,(A - B) + IP(B) + E_{ex}$$

admitting, of course that $AP\,(B^+)$ and $D\,(A - B)$ are known and that the excess energy can be neglected. Stevenson[9] has shown that for a large number of compounds $E_{ex} = 0$ if the ion corresponds to the fragment which, as a radical, has the lower ionization potential.

2   Reed

Bond energies can also be calculated in an indirect way using electron impact data in conjunction with thermochemical quantities. This is particularly useful when the ionization potentials of the fragments are not known. All that is necessary is to calculate the appearance potential of a certain fragment which can be obtained from several different molecules[9]. For instance, we will consider the case of the energy of the bond D (CH_3—H). The appearance potential will be used in thermochemical equations since, as we have already seen within certain limitations, it can be taken as the heat of the reaction.

From the following equations:

(i) $RCH_3 = R^+ + CH_3 + e - AP_1 (R^+)$

(ii) $RH = R^+ + H + e - AP_2 (R^+)$

(iii) $RH + CH_4 = RCH_3 + 2H + \Delta H_f^0 (CH_4) + \Delta H_f^0 (RH)$
$$- \Delta H_f^0 (RCH_3) - 2\Delta H_f^0 (H)$$

by addition of (i) and (ii) and subtraction of (iii) it is found

where $CH_4 = CH_3 + H - D (CH_3 - H)$

$$D (CH_3 - H) = AP_1 (R^+) - AP_2 (R^+) - \Delta H_f^0 (CH_4) - \Delta H_f^0 (RH)$$

$$+ \Delta H_f^0 (RCH_3) + 2\Delta H_f^0 (H).$$

CH_3–H bond energies calculated in several ways lead to the same values within the limits of experimental error which leads to the conclusion that in the present case the dissociation products are not formed in excited states. A similar conclusion is obtained in many other cases so that, apparently, the absence of excitation and kinetic energies of the dissociation products can be considered as being of common occurrence.

The mass spectrometric process can also be used to measure bond strength in positive ions[1]. For an ion AB^+, two processes of fission may be possible and obviously, two different values may be found according to the position of the charge in the fragmentation products. For each possible case the bond dissociation energy will be given by:

$$D_1 (A - B)^+ \leqslant AP (A^+) - IP (AB)$$

$$D_2 (A - B)^+ \leqslant AP (B^+) - IP (AB)$$

Again Stevenson's rule should be applied since the charge usually resides on the fragment with the lower ionization potential, and usually only one mode of fragmentation is observed.

3 CALCULATION OF HEATS OF FORMATION OF IONS

As has already been mentioned, the appearance potential of an ion can be considered as the heat of the reaction for the process involved in the formation of the ion so that the energetics of a reaction such as

$$AB + e \rightarrow A^+ + B + 2e$$

can be described in terms of the measured appearance potential in the following way:

$$AP\,(A^+) = \Delta H^o_{react} = \Delta H^o_f\,(A^+) + \Delta H^o_f\,(B) - \Delta H^o_f\,(AB) + E_{ex}$$

where E_{ex} is the excess energy involved. The heat of formation of the ion A^+ $[\Delta H^o_f(A^+)]$ can be calculated if the heats of formation of the other fragments and of the molecule are known and if, failing any information on E_{ex} it is assumed to be zero.

A criticism that can be made to this process of calculating heats of formation is that the temperature at which the measurements of appearance potentials are being performed is not usually known. Nevertheless the temperature effect is not large, and the assumption can be made that the conditions are those of a rare gas at $0\,^\circ C$ ($298\,^\circ K$). The heat of formation of the electron is taken as zero at all temperatures[13].

A very large number of data on heats of formation is available at present, a consequence of a publication by the National Bureau of Standards[13]. When experimental data are not available, there is a useful method of estimating heats of formation which has been presented by Franklin[14,15] using a group theoretical method. The method is based on the fact that thermodynamic quantities, such as heats of formation, may be represented as the sum of a set of values or group equivalents characteristic of the various groups that make up the molecule. Other methods have also been reported by Janz[16].

The information provided by the heats of formation of positive ions can be of great interest in understanding the structure of ions and molecules. An illustration of this is given by an example taken from the results of M. E. Wacks *et al.*[15] concerning salicylic acid and some of its esters. It was noted that several peaks corresponding to the same mass to charge ratio appeared. If the structure for an ion of a certain mass to charge ratio is common to all the compounds, then the heat of formation of the particular ion, calculated from ionization and appearance potential measurements should be the same. Experimental results show that this is so in the case of salicylic acid and some of its esters. Some interesting conclusions could then be drawn: as common ions, having the same heat of formation have the same structure, all the compounds considered follow similar fragmentation paths with the elimination of small neutral molecules such as water, carbon monoxide, ketene and methanol.

4   DETERMINATION OF ELECTRON AFFINITY

Closely related to the kind of calculations mentioned in connection with positive ions is the determination of electron affinities by means of the negative ions' mass spectra obtained by low energy electron impact.

Electron affinity can be defined as the exothermicity of the reaction

$$X + e \rightarrow X^-$$

considering both X and X^- on their ground states.

Negative ions in the gas phase may be formed by two different processes. One is an ion pair production process:

(i) $$AB + e \rightarrow A^+ + B^- + e$$

in which the ejected electron may carry some of the excess energy of the reaction. The conditions are similar to positive ion formation, and ionization efficiency curves for negative ions formed this way exhibit features which approach those of positive ions.

Another process of formation of negative ions is a resonance process which involves electron capture and can be accompanied (or not) by dissociation. Dissociation will occur when the captured electron has enough energy to promote a transition of the excited negative molecular ion above one of its dissociation asymptotes following the Franck–Condon principle:

(ii) $$AB + e \rightarrow AB^-$$

(iii) $$AB + e \rightarrow A + B^-$$

In both cases of the resonance capture process no electron is formed to carry away any excess energy, and because of this such a reaction will occur only if the energy of the bombarding electrons is within a narrow range. Figure 3 shows a classic example where both types of processes are observed.

The formation of negative ions by electron impact can be described by the Franck–Condon principle[17] as for positive ions.

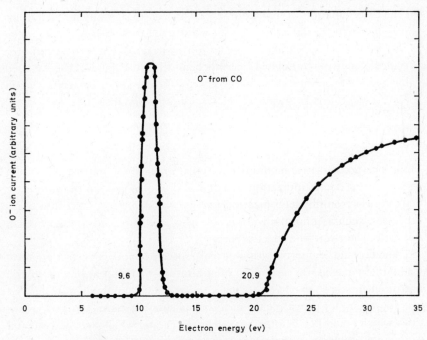

Figure 3   Ionization efficiency curve for O⁻ from CO[19]

The appearance potential of a negative ion is defined the same way as for positive ions. It can be related to the electron affinity by simple relationships.

For process (i)

$$AP(B^-) = D(A - B) + IP(A) + EA(B) + E_{ex}$$

where $EA(B)$ represents the electron affinity of B, E_{ex} stands for excess energy (kinetic or excitation energies or both of fragments) and the other quantities have the same meaning as previously mentioned.

For process (ii), electron affinity equals the appearance potential in the absence of excess energy, as seen from:

$$AP(AB^-) = EA(AB) + E_{ex}$$

For process (iii) the following relation holds:

$$AP(B^-) = D(A - B) + EA(AB) + E_{ex}$$

One of the main problems concerning negative ion production and the calculation of their appearance potential is the calibration of the energy scale for the determination of ionization efficiency curves. Usually a reaction involving ion-pair formation is used for which the excess energy can be taken as negligible. The appearance potential of the negative ion formed can then be calculated from tabulated thermochemical data[18] and the calculated value be used to calibrate the energy scale of the mass spectrometer. A good way of doing this is by means of the O^- ion obtained from CO.

5   CONCLUSION

The mass spectrometric methods have great applicability in the calculation of energies related to ionization and dissociation processes in the gas phase. Only the electron impact ionization process was mentioned here, but other processes, such as photoionization, are also used often advantageously in respect of the electron impact process.

The fact that the interpretation of mass spectra is not yet perfectly understood often means the accuracy of ionization and appearance potential measurements are low. Even so mass spectrometric methods have proved very important for providing the means of calculating thermodynamic quantities which could not be otherwise determined and which are fundamental to the interpretation of molecular structure.

It can be said that mass spectrometry has become an important technique to investigate the structure of matter in the ionized state.

References

1. J.H.Beynon, *Mass Spectrometry and its Applications to Organic Chemistry*, Elsevier Public. Co., London, 1960.
2. J.E.Collin, in *Mass Spectrometry* (R.I.Reed, ed.), Academic Press, London, 1965.
3. M.A.Haney and J.L.Franklin, *J. Chem. Soc.*, **48**, 4093, 1968.
4. G.Hertzeberg, *Electronic Spectra and Electronic Structure of Polyatomic Molecules*, Van Nostrand Co., London, 1967.
5. M.E.Wacks, *et al.*, presented at the 17th Annual Meeting of Mass Spectrometry and Allied Topics, ASTM E-14, Dallas, Texas, May, 1969 pps 442 *et sey.*
6. V.I.Vedeneyev *et al.*, *Bond Energies, Ionization Potentials and Electron Affinities*, Edward Arnold (publ.), London, 1966.
7. C.E.Melton, in *Mass Spectrometry of Organic Ions* (F.W.McLafferty, ed.), Academic Press, London, 1963.
8. R.E.Fox, W.M.Hickam, T.Kjeldaas, D.J.Grove, *Rev. Sci. Inst.*, **12**, 1101, 1955.
9. D.P.Stevenson, *J. Chem. Phys.*, **10**, 291, 1942.
10. M.A.A.Ferreira and M.E.Silva, to be published.
11. D.Beck, *Z. Phys.*, **160**, 406, 1960.

12. P.F.Knewstubb, *Mass Spectrometry and Ion Molecule Reactions*, University Press, Cambridge, 1969.
13. J.L.Franklin and J.G.Dillard, "Ionization Potentials, Appearance Potentials and Heats of Formation of Gaseous Positive Ions", *NSRDS-NBS* **26**, 1969.
14. J.L.Franklin, *Ind. Eng. Chem.*, **41**, 1070, 1949.
15. J.L.Franklin, *J. Chem. Phys.*, **21**, 2029, 1953.
16. G.J.Janz, *Estimation of Thermodynamic Properties of Organic Compounds*, Academic Press, New York, 1958.
17. H.S.W.Massey and E.S.Burhop, *Electronic and Ionic Impact Phenomena*, Oxford University Press, London, 1956.
18. R.W.Kiser, *Introduction to Mass Spectrometry and its Applications*, Prentice Hall, 1965.
19. H.D.Hagstrum and J.T.Tate, *Phys. Rev.*, **59**, 354, 1941.
20. W.A.Chupka, *J. Chem. Phys.*, **30**, 191, 1959.

Metastable ions

M. T. ROBERT-LOPES

Laboratório de Química, Faculdade de Ciencias, Lisboa 2, Portugal

According to their internal energy, the ions formed by electron impact in a mass spectrometer can be classified as:

stable-those which have insufficient excitation energy to decompose before collection;

unstable-those having enough energy to decompose before leaving the ionization chamber, and

metastable-those which possess intermediate energy decompose in transit.

The two first categories are observed in a mass spectrum as normal peaks. The last one gives rise to the so-called metastable peaks.

Aston[1] (1923) observed that diffuse peaks at non-integral mass numbers were found in mass spectra. In 1945, Hipple and Condon[2] reported that the non-integral masses appearing in the mass spectra of various hydrocarbons could be explained by the spontaneous dissociation of some of the ions into fragments of lighter mass after they had been accelerated and emerged from the ion gun. In 1946, Hipple, Fox and Condon[3] constructed an energy filter to make an energy analysis of the non integral masses in n-butane, butadiene and ethane. The values obtained agreed with those predicted on the basis of their arising from metastable ions assuming that there was a negligible release of kinetic energy during the dissociation.

If an ion of mass m_1 decomposes into an ion of mass m_2 and a neutral fragment of mass $(m_1 - m_2)$, and if this dissociation occurs after full acceleration, the ion m_2 appears at an apparent mass m^* given by

$$m^* = \frac{m_2^2}{m_1} \tag{1}$$

This relationship first given by Hipple and Condon[2], is valid for sector instruments and it holds also for $180°$ instruments.[2,4] In order that one can see how metastable peaks arise at the value m^* in the mass spectrum let the ion of mass m, formed in the ionizing region with zero kinetic energy fall through a potential difference of V_1 before decomposition into the ion of mass m_2 occurs. If there is only a very small release of internal energy during the dissociation, the neutral fragment moves with almost the same velocity as the ion of mass m_2 so this ion traverses the remainder of the accelerating voltage $V - V_1$ with a kinetic energy given by

$$\frac{1}{2} m_2 v^2 = \frac{m_2}{m_1} qV + q(V - V_1) \tag{2}$$

where v is the velocity of the ion of mass m_2 and q is the electronic charge. The ion of mass m_2 will enter the magnetic analyser with this kinetic energy after traversing the field free region (the case of a sector instrument). The radius with which it moves in the magnetic analyser is determined by

$$R = \frac{m_2 v}{Bq} \tag{3}$$

because in the magnetic field both centripetal and centrifugal forces, respectively Bqv, where B is the magnetic inductance and $m_2 v^2 / R$ are counterbalanced.

From (2) and (3) we get

$$R = \frac{(2V)^{1/2}}{Bq^{1/2}} \left\{ \frac{m_2^2}{m_1} \frac{V_1}{V} + m_2 \left(1 - \frac{V_1}{V} \right) \right\}^{1/2} \tag{4}$$

Therefore the ion of mass m_2 will be collected for the same combination of V and B as does a normal ion of mass m^* given by

$$m^* = \frac{m_2^2}{m_1} \frac{V_1}{V} + m_2 \left(1 - \frac{V_1}{V} \right) \tag{5}$$

Nevertheless a very small number of ions formed at large values of $(V - V_1)$ will be collected, because of the directional focusing properties of the sector instrument and the discrimination due to a very small release of internal energy[5]. Thus the mass of the ions m_2 formed by a metastable decomposition is usually found at the value given by the expression (1).

In any kind of magnetic deflexion instrument we have metastable decompositions along the whole tube of the mass spectrometer starting immediately after the creation of metastable ions in the ion source until they reach the collector. This fact gives rise to a spread in the range of the times over which these decompositions can be observed. Metastable ions have a life time between 10^{-6} and 10^{-5} sec.

The intensity of the metastable peak varies linearly with sample pressure and it is mostly due to the decompositions occurring before the magnetic analyser in the case of sector instruments. The daughter ions formed in any of the other regions will give a very small contribution to the intensity of the metastable peak. They will affect the focusing of the ions formed before the magnetic analyser which appear at the value m^* of the mass scale, as we shall see below. Usually there is a small release of internal energy causing the metastable peaks to be broader than normal peaks. If we alter the residence time of the metastable ion within the ionization chamber, the intensity of the metastable peak changes relative to the intensity of the normal peaks in the mass spectrum.[5] This can be done by varying the ion repeller potential relative to the ionization chamber.

The intensity of the metastable peaks is increased when the sector instruments are operated with the lowest resolution rather than with the highest.[5]

In the case of the 180° instrument the intensity of the metastable peak is due to the decompositions taking place in a short path length[4] within the magnetic analyser as shown in Figure 1.

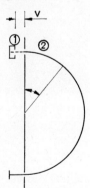

Figure 1   Schematic diagram of 180° single focusing mass spectrometer

The relative sensitivities of single and double focusing geometries in the collection of daughter ions from metastable transitions were discussed by Barber and Elliott[6]. They concluded that the intensity of the metastable peaks is dependent mainly on the geometry of the magnetic analyser and that the double focusing principle is not relevant. Thus spectra of the same compound obtained with the same resolution for example with the instruments shown in Figure 2 and Figure 3 will show the same metastable peaks

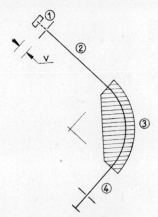

Figure 2   Schematic diagram of 90° sector magnet single focusing mass spectrometer

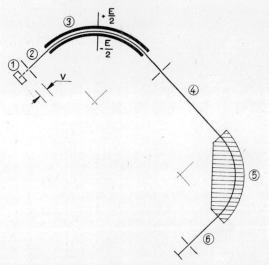

Figure 3   Schematic diagram of Nier–Johnson double focusing mass spectrometer

with about the same intensity, because the geometry of the magnetic analyser is the same for both mass spectrometers.

The mass m^* of a metastable peak should be measured accurately. In some spectra we may have metastable peaks intense enough to make accurate mass measurements, i.e. in the aniline spectrum[7]

$$m_1 \qquad m_2 \, (m_1 - m_2) \qquad\qquad m^*$$

$$93^{+\cdot}-66^{+\cdot} + 27 \qquad\qquad\qquad 46.8$$

$$C_6H_7N^{+\cdot}-C_5H_6^{+\cdot} + HCN \qquad 46.876$$

$$C_6H_7N^{+\cdot}-C_4H_4N^+ + C_2H_3 \quad 46.858$$

Nevertheless in most of the cases they are of very low intensity, less than 1% of the most intense peak in the spectrum and it becomes impracticable to do exact mass measurements. In other cases they may not be resolved due to overlap of normal peaks.

Thus the value for m^* is obtained by the trial-and-error method.

When working with a double focusing mass-spectrometer it is possible to identify metastable transitions accurately by introducing circuit modifications.

In double focusing mass-spectrometers of the Nier–Johnson type, the ratio of the accelerating voltage V_0 to the electrostatic analyser voltage E_0 is such that single charged ions of kinetic energy V_0q are transmitted.

After dissociation the kinetic energy of the daughter ions is a fraction of V_0q:

$$\frac{1}{2} m_2 v^2 = \frac{m_2}{m_1} V_0 q \tag{6}$$

and if E_0 is held constant, it is sufficient to increase the accelerating voltage to the value

$$V = V_0 \frac{m_1}{m_2} \tag{7}$$

so that the daughter ions are transmitted.

The equations for the ion m_1 moving in the electrostatic analyser are

$$\frac{m_1 v^2}{R} = qE_0 \tag{8}$$

and

$$\tfrac{1}{2} m_1 v^2 = qV_0 \tag{9}$$

where R is the radius of the electrostatic analyser.

We have from (8) and (9)

$$R \doteq \frac{2V_0}{E_0} \tag{10}$$

for normal ions, but for ions decomposing inside the analyser we have

$$R_1 = \frac{2V_0 m_2}{m_1 E_0} \tag{11}$$

so that when the accelerating voltage is changed for a value V,

$$R = \frac{2(V_0 + \Delta V)}{E_0} \cdot \frac{m_2}{m_1} = \frac{2V}{E_0} \tag{12}$$

and normal ions are no longer transmitted.

Jennings[8] employing Nier–Johnson geometry and Futtrell and coworkers[9] with Mattauch–Herzog ion optics have used this technique which prevents the passage through the mass spectrometer of all ions that have acquired the full accelerating energy (normal ions) and allows only daughter ions which have been formed in the field free region preceding the electrostatic analyser (2 Figures 3, 4) to be recorded. Osberghaus[10] and Ottinger[10,11] measured fragmentation energies in hydrocarbons using an arrangement closely similar to this one. Reed and Robert-Lopes[12] applied this method to map metastable ion decompositions in a series of monoterpenes.

The magnetic field is set to focus daughter ions of mass m_2 at the collector.

Keeping the electrostatic analyser voltage E_0 constant the accelerating voltage is then increased from V_0 to V so that only daughter ions formed before the electrostatic analyser will be collected. From the measured values of V and knowing the mass of the daughter ion m_2, the expression (7) will give us the mass value of the different precursors m_1. The metastable peaks suffer no interference from normal peaks and have an increased intensity because the electron multiplier voltage can be much higher than when the instrument is functioning normally.

Nevertheless we have to be careful when using this method, because in the particular case of an A.E.I. M.S.9 mass spectrometer it is possible to vary E/V by 1.3%, and still have normal ions with energies within $\pm0.66\%$ passing through the analyser. This fact is disadvantageous to the above method because when the daughter ion has several precursor ions differing in mass by less than 1.3% these can be scarcely resolved, as shown by Beynon.[13]

Very recently, Beynon *et al.*[14] considered the possibility of increasing the energy resolving power of the elctrostatic analyser by narrowing the width of the monitor slit. They substituted the standard 0.2 in width monitor slit by a variable one. Thus they observed the metastable peaks corresponding to some competing decomposition reactions now completely resolved.

Daly, McCormick and Powell[15] built a special scintillator counter to detect daughter ions of metastable transitions occurring both in the field free region before the magnetic analyser and before the collector. It can be used with both single and double focusing mass spectrometers. In a single sweep of the magnetic field the daughter ions from metastable decompositions are detected in a mass spectrum, thus being unequivocally identified.

So far, the release of internal energy during the dissociation of the metastable ion has been neglected. If we take into account, the release of internal energy can appear as kinetic energy of the fragments and can strongly affect the shape and the width of the corresponding metastable peak.

When looking carefully at a mass spectrum one might observe a variety of shapes and widths. Indeed, although the shapes of most of the peaks is roughly gaussian, some other shape like "continuum", "triangular", "flat-topped" and "dished topped" can be found. These shapes may extend over a range of several mass numbers. Sometimes they are difficult to define because of the overlapping with normal peaks. In any case the peak should be centered about a mass value given by[16]

$$m^* = \frac{m_2^2}{m_1}\left[1 + \frac{\mu T}{qV}\right] \qquad (13)$$

where μ is the reduced mass $(m_1 - m_2)/m_2$ and T the mean energy released.

This value is very close to m^* from expression (1) which was deduced without considering the release of internal energy.

A "continuum" can be due to: (a) a metastable ion decomposition between the plane of the electron beam and the exit slit in single focusing 180° and sector magnet instruments, (**1** Figure 1 and, **2**). The daughter ions appear at masses lying apparently between m_2 and m_2^2/m_1, giving rise to a tail of the metastable peak toward higher masses.[4,6] In double focusing instruments only those ions, formed in that region, (**1** Figures 3, 4) possessing adequate kinetic energy to pass through the electrostatic analyser will reach the collector. So in general ions formed in that region are not collected.[6] (b) If the decomposition occurs in the electrostatic analyser of the double focusing instrument (3 Figures 3 and 4) some of the ions will not be transmitted because the path of these ions has a radius of curvature different from that of the electrostatic analyser, as seen above; other ions will pass through but will contribute to a long-tailed continuum from the metastable peak to the corresponding normal peak.[6,17] (c) If decompositions take place within the magnetic field of a 180° single focusing instrument[14] [18] (**2** Figure 1) or within

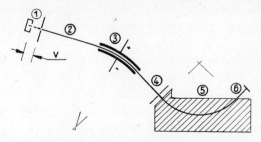

Figure 4   Schematic diagram of Mattauch–Herzog double focusing
mass spectrometer

a sector magnetic field (Figure 2) as considered by Beynon and Fontaine.[19] In any case the resulting daughter ions are transmitted only if they obey Coggeshall's mathematical condition[4] which is the reason why in a 180° instrument there is only a short path length corresponding to observable metastable peaks. In a sector instrument the daughter ions resulting from decomposition within the magnetic field will be collected when the decomposition takes place either very near the beginning, or very near the end of the magnetic field. Daughter ions formed at any of the other points will originate a continuum from m_2^2/m_1 to m_1, unless the mass change is about 1%. In this latter case they all are collected. (d) If decompositions occur before the collector (**6** Figures 2, 3 and 4) they will give rise to a tail on each side of the m_1 peak.[19] All these cases contribute to the distortion of the shape of the metastable peak.

Metastable peaks much broader than usual with a relatively flat-top were first observed by Newton and Schiamanna.[20] Beynon, Saunders and Williams[16] reported metastable peaks abnormally wide. In the spectrum of *o*-nitrophenol they found a well-defined "flat-topped" peak. Jennings and Higgins[21] observed a "dished-topped" peak in the decomposition of doubly charged ions. A broad triangular peak was observed by Reed and Robert-Lopes[22] in the spectra of some monoterpenes.

The broadening of the width of a "flat-topped" peak was shown to be related to the energy released during the decomposition by the following expression

$$d = (4m_2^2/m_1) \sqrt{\mu T/qV} \qquad (14)$$

where d is the width of the metastable peak in atomic mass units corresponding strictly to the flat portion as seen in Figure 5. Energy release was considered only along the direction of motion of the ions.

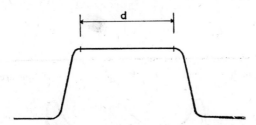

Figure 5   Flat-topped metastable peak

Elliot[19] and Flowers[23] considered the possibility of the energy release in a metastable decomposition producing a velocity component in the daughter ion marking an angle with the direction of the main ion path. They developed equations related to the geometry of a MS9 mass spectrometer. The trajectories of daughter ions from metastable decompositions occurring at different positions in the field free region in front of the magnetic analyser were calculated. Flowers deduced equation (15).

$$2\Delta m^*/m^* = \Delta v \left\{ \cos \theta \, (2 - \Delta v \cos \theta/v) \pm (2.79 - 3.95d/55) \sin \theta \right\}/v \qquad (15)$$

where Δm^* is the shift from the m^* mass value of the daughter ion from a metastable decomposition occurring at a distance $d$cm from the beginning of the field free region, v is the velocity of the metastable ion due to the accelerating voltage (V), Δv is the velocity increment resulting from any kinetic energy released in the decomposition of the metastable ion, θ is the angle of Δv with the main ion beam, $\Delta v \cos \theta$ being the velocity vector adding to or subtracting from the overall velocity. A "dish topped" peak was predicted from these calculations (Figure 7). Momigny[19] and Flowers[23] discussed the exact position at which metastable peaks should appear on the mass scale.

The variations of the peak shape with different accelerating voltages were studied by McLafferty *et al.*[24] They observed that the usual gaussian shaped metastable peak could change the round top to a flat or dish-top when the accelerating voltage was decreased.

3   Reed

Beynon and Fontaine[19] studied, in great detail, the shape of metastable peaks. They calculated the discrimination effects against daughter ions formed at any point along the flight path tube. They studied, theoretically,

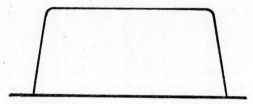

Figure 6   Metastable peak with a flat top, assuming all ions are collected

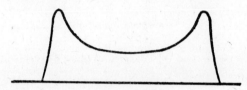

Figure 7   Metastable peak with a dish top, assuming discrimination by a collector slit of finite length

the effects of finite beam dimensions, angular spreading and kinetic energy release during the decomposition. They found what the computed peak shapes, taking into account release of energy not only along the direction of motion but along other directions such as the length of the slit, would be like (Figures 6 and 7).

Generally, in practice, the slopes of the sides of all broad metastable peaks are less than in the normal ion peaks. Theoretically[19] they should be as steep as those of the normal peaks. This is explained by the fact that the decomposition of metastable ions does not always occur with the same amount of energy release; there must be a variable amount of energy liberated during dissociation.[7] Thus the different shapes for the abnormally wide metastable peaks depend on the shape of the energy distribution function.

Beynon, Hopkinson and Lester[25] using a simplified version of quasi-equilibrium theory derived from the peak width the value for the number of effective oscillators. According to the quasi-equilibrium theory[26] the simplest form of the rate constant for the processes in a mass spectrometer is given by

$$K = v \left(\frac{E - E_0}{E} \right)^{s-1}$$

where v is the frequency factor, E the internal energy, E_0 the activation energy and s the effective number of oscillators. They considered the total release of kinetic energy T in the case of a metastable ion and related this value with the

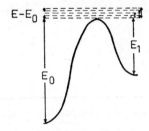

Figure 8 Schematic representation of the energy relationship in an energetic metastable decomposition

forward and reverse activation energies (Figure 8) respectively E_0 and E_1. They deduced the following expression

$$T = E_1 + E_0 \left(1 - 2^{-1/(s-2)}\right) \frac{(K/v)^{1/(s-1)}}{1 - (K/v)^{1/(s-1)}} \tag{16}$$

From (15), E_1 and E_0 can be evaluated for given values of s and v, two values of K with the corresponding values of T taken at half-height.

As was mentioned above the m^* value must be measured accurately in order that we can determine the corresponding transition $m_1^+ \rightarrow m_2^+ + (m_1 - m_2)$. Beynon, Saunders and Williams[27] computed metastable tables where m_1 and m_2 values can be found. Each pair of mass values helps us to understand how a molecular ion or fragment ion decomposes in one step. The neutral fragment is most likely lost as an entity.[5] Thus the presence of the metastable peak can provide valuable evidence about the arrangement of atoms in a molecule. When we have more than one compound in a mass spectrum the existence of metastable peaks becomes extremely important as each pair of ions comes from a unique compound. Beynon[7] showed how by means of metastable peaks an impurity was isolated from a reaction product which enhanced their utility in the analysis of mixtures. The molecular formula of the impurity was established by accurate mass measurement and by means of a series of metastable peaks a fragmentation pathway was determined. The structure of the impurity could be concluded by analogy with the structure of the main reaction product.

However metastable peaks can be prejudicial in quantitative analysis of mixtures of known components when they interfere with the measurement of the exact height of normal peaks. In such a case the use of metastable suppressors is indicated, which can be achieved by applying a repelling potential to the collector.[5] This procedure may give additional information about the kinetic energy of the daughter ions from metastable decompositions, which can thus be measured accurately. By means of metastable peaks, mechanisms of fragmentation of known organic compounds are elucidated.[28] We can get a complete fragmentation pathway where all ions come from metastable transitions.

Jennings[29] observed a two step decomposition process during transit in the field free region in front of the electrostatic analyser using the technique mentioned above which prevents normal ions from passing through the electrostatic analyser.

$$C_7H_7^+ \xrightarrow{46.4} C_5^+H_5 \xrightarrow{23.45} C_3H_3^+$$
$$\searrow 16.8$$
$$C_3H_3^+$$

Measurements of large metastable peak widths have been used to give thermodynamic information about ion fragmentation processes.[30,31] In the particular case of doubly charged ions, intercharge distances can be estimated together with thermodynamic information which enables us to deduce the structures of these ions.[32]

When decomposition of a metastable doubly-charged ion into two singly-charged ions occurs, there is a release of kinetic energy due to repulsion between the two charges. Depending on the stability of the new fragment ions further energy release may or may not take place. In the mass spectrum of benzene the metastable decomposition (17)

$$C_6H_6^{2+} \rightarrow C_5H_3^+ + CH_3^+ \tag{17}$$

gave rise to two "dish-topped" metastable peaks and it was shown by Jennings[21] that there was an energy release of 2.8 eV during decomposition. This energy release was calculated from the width of the metastable peak by the process referred to previously. Supposing that the energy release of 2.8 eV comes strictly from the charge separation in the metastable doubly charged ion, then the intercharge distance is 5.14 Å as calculated from (17)

$$T = \int_r^\infty \frac{q^2}{r^2} \, dr \tag{18}$$

where q is the electronic charge and r the initial intercharge distance. As there are two positive charges repelling each other, T represents the work of separation of these two charges from their initial distance apart to infinity. If only part of the energy released arises from the charge separation, then the intercharge distance will be greater than 5.14 Å. It was proposed that the loss of two electrons from benzene should be visualized as leading to an opening of the ring and could involve electron rearrangement in the following manner

Shannon and McLafferty[33] studied ions of elemental composition $C_2H_5O^+$ from spectra of a wide variety of compounds on the basis of the mass, relative abundance and shape of the peaks resulting from metastable decompositions of these ions. They found four distinct structural types. Further studies[34] took into account the effect of the number of vibrational degrees of freedom in the molecular ion. It was found that the logarithm of the abundance of the metastable ions $C_2H_5O^+$ is inversely proportional to the number of vibrational degrees of freedom in the parent molecular ion.

Metastable peaks from collision induced metastable decompositions are reported[35] as having relative abundances high enough and appearing to be nearly independent of the energy of their precursor ions. In such processes the structure of an ion could be characterized by the abundance of a single metastable peak.

Metastable ion abundances may characterize ion decomposition mechanisms.[36] The argument consists in considering a process not involving rearrangement when an abundant normal ion peak has not a correspondingly abundant metastable ion peak. By use of collisional activation energy,[37] rearrangement reactions can be reduced or eliminated in mass spectra which becomes very important in structural studies. Very recently McLafferty *et al.*[38] pointed out the importance of metastable transitions to indicate the energetically most-favoured processes.

Metastable transitions of a wide variety of compounds were studied and they found that often the most abundant ions arise from reaction pathways

$$\tag{19}$$

not recognized from normal peaks. As an example they showed the formation of a five-membered ring product (reaction 19) in a low activation energy rearrangement reaction. For $CH_3CH_2CH_2CD_2COCD_3$, the metastable peak corresponding to the loss of CH_3 is the most abundant from the molecular ion and is about six times greater than that, corresponding to the loss of CD_3. The appearance potential of $(m\text{-}CH_3)^+$ is slightly higher than that of the molecular ion, and despite the low activation energy for this reaction a very low abundance of the corresponding normal daughter ions is observed.

Metastable ion decompositions have mostly been studied in magnetic deflexion instruments. Metastable transitions in field ionization mass spectrometers were reported by Beckey.[39] Very recently with the R.M.I. Hitachi double-focusing instruments Struck[40] has shown that one can recognize in which field free region metastable transitions take place. In these instruments the field free regions are extended and there is a long field free region between the source and the electrostatic analyser. More recently, the use of radio-frequency mass spectrometers has been proposed for the study of these transitions, namely the time of flight (t-o-f) mass spectrometer.[41-45] The addition of a retarding region before the collector made possible the detection of some metastable transitions which could not be seen in magnetic sector instruments. Lately,[46] an instrument modification to defocus interfering normal ions associated with a computer method of analysis makes it possible to calculate the masses of the daughter ions accurately.

Finally, metastable transitions are of great interest in theoretical studies of mass spectra as will be seen in the next chapter by G. R. Lester.

References

1. Aston, F. W. (1923) *Phil. Mag.* **45**, 940.
2. Hipple, J. A. and Condon, E. U. (1945) *Phys. Rev.* **68**, 54.
3. Hipple, J. A., Fox, R. E. and Condon, E. U. (1946) *Phys. Rev.* **69**, 347.
4. Coggeshall, N. D. (1962) *J. Chem. Phys.* **37**, 2167.
5. Beynon, J. H. (1960) *Mass Spectrometry and its Application to Organic Chemistry*, p. 251, Elsevier, London.
6. Barber, M. and Elliott, R. M. (1964) *Proceedings of the 12th ASTM E-14 Conf.*, Montreal, p. 150.
7. Beynon, J. H. (1968) *Advances in Mass Spectrometry*, vol. 4, editor E. Kendrick, p. 123, the Institute of Petroleum.
8. Jennings, K. R. (1965) *J. Chem. Phys.* **43**, 4176.
9. Futrell, J. H., Ryan, K. R. and Sieck, L. W. (1965) *J. Chem. Phys.* **43**, 1832.
10. Öttinger, Ch. and Osberghaus, O. (1965) *Physics Letters* **17**, 269.

11. Öttinger, Ch. (1965) *Zeit. Naturf.* **20a**, 1229.
12. Reed, R.I. and Robert-Lopes, M.T., work to be published.
13. Beynon, J.H. and Fontaine, A.E. (1967) *Some Newer Physical Methods in Structural Chemistry*, editors R.Bonnett and J.G.Davis, United Trade Press, London, p. 111.
14. Beynon, J.H., Fontaine, A.E., Hopkinson, J.A. and Williams, A.E. (1969) *Int. J. Mass Spect. Ion Phys.* **3**, 143.
15. Daly, N.R., McCormick, A. and Powell, R.E. (1968) *Rev. Sci. Inst.* **39**, 1163.
16. Beynon, J.H., Saunders, R.A. and Williams, A.E. (1965) *Zeit. Naturf.* **20a**, 180.
17. Beynon, J.H., Saunders, R.A. and Williams, A.E. (1964) *Nature* **204**, 67.
18. Newton, A.S. (1966) *J. Chem. Phys.* **44**, 4015.
19. Beynon, J.H. and Fontaine, A.E. (1967) *Zeit. Naturf.* **22a**, 334.
20. Newton, A.S. and Sciamanna, A.F. (1964) *J. Chem. Phys.* **40**, 718.
21. Higgins, W. and Jennings, K.R. (1966) *Trans. Faraday Soc.* **62**, 97.
22. Reed, R.I. and Robert-Lopes, M.T. (1966) *Proceedings of the 14th ASTM Conference*, Dallas, p. 445.
23. Flowers, M.C. (1965) *Chem. Comm.*, 235.
24. Shannon, T.W., McLafferty, F.W. and McKinney, C.R. (1966) *Chem. Comm.*, 478.
25. Beynon, J.H., Hopkinson, J.A. and Lester, G.R. (1968) *J. Mass Spectrometry and Ion Phys.* **1**, 343.
26. Rosenstock, H.M., Wallenstein, M.B., Wahrhaftig, A.L. and Eyring, H. (1952) *Proc. Natl. Acad. Sci. U.S.* **38**, 667.
27. Beynon, J.H., Saunders, R.A. and Williams, A.E. (1966) *Tables of Metastable Transitions*, Elsevier, Amsterdam.
28. Biemann, K. (1962) *Mass Spectrometry*, McGraw-Hill Book Co. New York, p. 153.
29. Jennings, K.R. (1966) *Chem. Comm.*, 283.
30. Beynon, J.H., Fontaine, A.E. and Lester, G.R. (1968) *J. Mass Spect. Ion Phys.*, **1**, 1.
31. Bursey, M.M. and McLafferty, F.W. (1966) *J. Am. Chem. Soc.*, **88**, 5023.
32. Beynon, J.H. and Fontaine, A.E. (1966) *Chem. Comm.*, 717.
33. Shannon, T.W. and McLafferty, F.W. (1966) *J. Am. Chem. Soc.* **88**, 5021.
34. McLafferty, F.W. and Pike, W.T. (1967) *J. Am. Chem. Soc.* **89**, 5951–5954.
35. Haddon, W.F. and McLafferty, F.W. (1968) *J. Am. Chem. Soc.* **90**, 4745.
36. McLafferty, F.W. and Fairweather, R.B. (1968) *J. Am. Chem. Soc.* **90**, 59, 15.
37. McLafferty, F.W. and Schuddemage, H.D.R. (1969) *J. Am. Chem. Soc.* **91**, 1866.
38. McLafferty, F.W., McAdoo, D.J. and Smith, J.S. (1969) *J. Am. Chem. Soc.* **91**, 5400.
39. Beckey, H.D. (1965) *Mass Spectrometry*, A N.A.T.O. Advanced Study Institute on theory, design and applications. editor R.I.Reed, Academic Press.
40. Struck, A.H. and Major, H.W. jr., (1969) *Proceedings of the 17th ASTM E-14 Conference*, Dallas, p. 102.
41. Ferguson, R.E., McCulloh, K.E. and Rosenstock, H.M. (1965) *J. Chem. Phys.* **42**, 100.
42. McLafferty, F.W., Golke, R.S. and Golesworthy (1964) *Proceedings of the 12th ASTM Conference on Mass Spectrometry*, Montreal.
43. Hunt, W.W. Jr., Huffman, R.E. and McGee, K.E. (1964) *Rev. Sci. Instrum.* **35**, 82.
44. Hunt, W.W. Jr., Huffman, R.E., Saari, J., Wassel, G., Betts, J.F., Paufve, E.H., Wyess, W. and Fluegge, R.A. (1964) *ibid.* **35**, 88.
45. Dugger, D.L. and Kiser, R.W. (1967) *J. Chem. Phys.* **47**, 5054.
46. Haddon, W.F. and McLafferty, F.W. (1969) *Anal. Chem.* **41**, 31.

Recent studies of metastable ions and some mechanistic interpretations

G. R. LESTER

Imperial Chemical Industries Limited, England

INTRODUCTION

The possibility of deriving thermochemical data from metastable peaks in mass spectra seems first to have been realized by Beynon *et al.*[1], and it has since become evident that the same technique may be applied to a wider study of potential energy surfaces[2,3]. Unsuspected features such as "inversion" effects presented by deuterated derivatives[2], can be readily explored using this technique. It is not simply a question of the negotiation of a col in an energy surface; of considerable importance is the "kinetic shift" or excess energy required to yield an ion of appropriate lifetime to appear in the field-free region, and the redistribution of such energy excess in the product species. In particular, it is instructive to consider properties such as the shape of the potential energy surface that determine whether vibrational quanta can emerge as excess kinetic energy or merely apportioned between charged and neutral fragments. If the latter is the case, its effect disappears from thermochemical energy balances. The sharpness of a metastable peak can be shown to provide detailed information on the number of effective oscillators and of forward and backward activation energies.

Much of this article will be concerned with anomalous effects in metastables. Whilst some observations are apparently anomalous, it is usually possible to provide acceptable or plausible mechanistic interpretation of some kind. However quite extraordinarily, some results even appear incompatible with the laws of mechanics! For example, in the anomalous mass effect observed in the methanol mass spectrum, it is almost as though a fragment ion can spontaneously increase its own momentum in the field-free region, a sort of "bootstrap" effect. This is, of course, patently absurd; some species of stray field or obscure collisional phenomenon is doubtless responsible, otherwise momentum is necessarily conserved. But so far explanations along such lines suffer from the serious objection that the calculated cross-sections are apparently not of the right order of magnitude.

It seems clear that this field of metastable ions is rich in unexpected situations calling for ad hoc explanations, as well as for new theories of energy distribution and transfer in polyatomic ions. It will be the purpose of this article to examine some of the lines of approach that have recently seemed worth pursuing. Whilst mainly classical considerations seem to suffice, it is interesting to note that some wave-mechanical ideas have close classical analogues. This seems to provide an indication of how contemporary theory might be usefully developed.

1   EXCITATION BY ELECTRON IMPACT

Basic to the discussion of any dissociation mechanism is that of the initial activation by electron impact. There are two essential aspects, namely the electron excitation or ionization and the resulting vibrational effects. The latter are commonly discussed in terms of the Franck–Condon principle and can be considered separately in view of the mass disparity of electron and atomic nuclei (Born–Oppenheimer principle). This separation into electric and vibrational factors in the wave function is very widely adopted in most discussions of molecular properties. It is hardly necessary to elaborate on the vibrational aspects because these have been discussed extensively in various presentations.

The problem of electronic excitation is on the whole somewhat more complicated, and has received scant attention so far. An attempt has been made[4] to draw attention to a number of parallels with related molecular phenomena, such as the polarization effect of a molecule when subject to a ligand field or to a travelling electromagnetic wave. On the whole the latter effects though not entirely dissimilar in principle, are less complex and can be handled by readily accessible theoretical procedures. It is of interest to try to discern under what situations electron impact may be discussed in a similar tractable form. The requirements essentially one (i) that a steady state of initial excitation is reached in which the molecule is subject to bombardment by the electron beam for a time long relative to the transition times, spontaneous and induced; (ii) there is a sufficient delay between the setting up of the initially prepared state and the subsequent dissociation for the resulting initial properties to be directly relevant; (iii) the impact electrons have so much more energy than the bound electrons that to a large extent the path of the impact electron is not widely disturbed by the perturbing field. For 70 eV electrons, this is substantially true for excitations requiring less than about 10 eV.

Instead of a time-dependent formalism it is then permissible to use the stationary state method, transient effects due to switching on the electron beam being unimportant. It is not possible to normalize the wave function e^{ikz} of a free electron and physical interpretation as an electron flux is the only permissible one. Because the impact electrons are of relatively high energy the scattered electrons are strongly concentrated in the initial beam direction, exactly analogous to the classical case of Rutherford scattering where the angular dependence is the familiar $\operatorname{cosec}^4 \frac{1}{2}\theta$. In this connection Born[5] has drawn an analogy with the vibration of air in an organ pipe; if a key is depressed for a sufficient duration a proper recognizable tone results but a sharp staccato would produce a disagreeable distorted sound. We may therefore regard the mass spectrometer as an instrument on which recognisable tunes can be produced when operated in the correct manner.

Let q represent a complete set of eigenvalues for the states of the undisturbed parent molecule, so that a set of wave-functions $\phi_0(q), \phi_1(q), \ldots, \phi_n(q), \ldots$ exists with energies $E_0, E_1, \ldots, E_n, \ldots$ Let $\mathbf{r}$ be a position vector of the impact electron and $V(\mathbf{r}, q)$ the perturbation potential due to its presence.

Then the modified wave equation for steady state conditions is:

$$\left\{ \frac{-\hbar^2}{2m} \nabla^2 + H(q) + V(\mathbf{r}, q) \right\} \psi(\mathbf{r}, q) = E\psi(\mathbf{r}, q)$$

If $\mathbf{k}_0$ is the wave vector of the impact electron directed along the z direction, its wave function is $e^{i\mathbf{k}_0 \cdot \mathbf{r}}$ or $e^{ik_0 z}$. The total energy, which must be conserved, is $E = (1/2m) \hbar^2 k_0^2 + E_0$ corresponding to that of a free electron with accurately defined momentum and the ground state of the molecule. As $V(\mathbf{r}, q) \to 0$, clearly $\psi(\mathbf{r}, q)$ must $\to e^{i\mathbf{k}_0 \cdot \mathbf{r}} \phi_0(q)$.

For the perturbed wave-function of the whole system in a stationary state, we can write

$$\psi\,(\mathbf{r},\,q) = \sum_m \phi_m(q)\,\psi_m(\mathbf{r})$$

$$\therefore \sum_m \phi_m\,(q) \left\{\nabla^2 + \frac{2m}{\hbar^2}\,(E - E_m)\right\} \psi_m(\mathbf{r}) = \frac{2m}{\hbar^2}\,V\,(\mathbf{r},\,q)\,\psi\,(\mathbf{r},\,q)$$

$$= U\,(\mathbf{r},\,q)\,\psi\,(\mathbf{r},\,q)$$

where

$$U\,(\mathbf{r},\,q) = \frac{2m}{\hbar^2}\,\psi\,(\mathbf{r},\,q)$$

Since by orthonormality $\int \phi_m(q)\,\phi_s(q)\,dq = \delta_{ms}$

$$(\nabla^2 + k_s^2)\,\psi_s(\mathbf{r}) = \int dq\,\bar{\phi}_s\,(q)\,U\,(\mathbf{r},\,q)\,\psi\,(\mathbf{r},\,q) = F_s(\mathbf{r})$$

This equation is of a well-known type. For example, if $k_s \equiv 0$ and $F_s(\mathbf{r})$ is the Dirac δ function it corresponds to the Poisson equation for a point charge

$$\nabla^2 V = -\delta(\mathbf{r})$$

with coulomb type solution $V = 1/4\Pi r$.

Often it will suffice to regard the co-ordinates $\mathbf{r}_1$ as referring to one bound electron, for it is primarily one electron (e.g. a non-bonding electron in a hetero atom) that interacts with the impact electron. Under these conditions $V\,(\mathbf{r},\,q)$ simplifies to $V\,(\mathbf{r},\,\mathbf{r}_1) = e^2/|\mathbf{r} - \mathbf{r}_1|$
where

$$|\mathbf{r} - \mathbf{r}_1| = \sqrt{r^2 + r_1^2 - 2rr_1 \cos\theta_1} = r\,\sqrt{1 + \left(\frac{r_1^2}{r}\right) - 2\,\frac{r_1}{r}\,\cos\theta_1}$$

$$V = \frac{e^2}{r}\left[1 - \frac{1}{2}\left(\frac{r_1}{r}\right)^2 + \frac{r_1}{r}\,\cos\theta_1 + \frac{3}{2}\left(\frac{r_1^2}{r}\right)\cos^2\theta_1\cdots\right]$$

Thus $F_s(r)$ involves components like

$$e^2/r^2 \int \bar{\phi}_s(\mathbf{r}_1)\,r_1 \cos\theta_1 \psi\,(\mathbf{r}\,\mathbf{r}_1)\,d\tau_1 = \sum_m (e^2/r)\,\psi_m(\mathbf{r}) \int \bar{\phi}_s(\mathbf{r}_1)r_1 \cos\theta_1 \phi_m(\mathbf{r}_1)\,d\tau_1$$

These depend upon the same transition moment integrals as those familiar in excitation by a beam of light. But in addition terms of higher order emerge owing to their presence in V.

The final evaluation leads to a scattered electron beam of varying intensities in assigned directions with respect to the initial beam direction. It is important to note that the scattered beam cannot have sharply defined momentum. That is to say, it is not the case that certain specific scattering directions will be associated with corresponding excitation effects, polarized accordingly. The maximum ascertainable knowledge of the scattering process relates to the overall wave function which has components corresponding to all arbitrary scattering angles. Thus, if the initial beam has symmetry about one direction (the z direction), for a single atom we expect to find an equivalent symmetry in the perturbed state.

It is just as though an equivalent strong electric field had been applied along the same z-direction. Lateral distortion may be ignored because the whole phenomenon must retain a symmetry axis.

2 STATIONARY STATES FOR ELECTRON IMPACT AND RADIATION FIELDS

In an electron impact process we are concerned with a strong field interaction in the terminology of ligand field theory.[6] The potential field due to the impact is comparable in magnitude with that responsible for internal cohesion of the system. In a classical model this would imply that accelerations and therefore time scales are similar for both—thus, $\tau \sim 10^{-14}$ sec. For spontaneous emission $\tau \sim 10^{-8}$ sec (the mean lifetime of an excited state before it emits radiation) which implies much lower rates of spontaneous electron disturbance. In the radiational field case, transition probabilities A and B[7] would correspondingly satisfy

$$A \ll Bu$$

where u = radiational density.

In the case of a radiation field the population of excited states is usually extremely small compared to the ground state, thus $Bu\, n_1 = (A + Bu)\, n_2$ in thermal equilibrium or

$$\frac{n_1}{n_2} = \frac{A + Bu}{Bu} \gg 1$$

where $A \gg Bu$.

This preponderance of spontaneous over stimulated emission is typical of radiational activation and in electron impact the opposite condition applies.

If $A \ll Bu$

$$\frac{n_1}{n_2} \approx 1$$

so that ground and excited states are equally occupied. What is the corresponding condition for a steady electron beam?

If $H\Psi = E\Psi$ is the wave equation for the isolated molecule, then the
corresponding equation for the steady state under electron-impact is

$$(H + U)\Psi = E^1\Psi$$

where U represents the effective potential field generated by distortion of
the electron beam.

Let U, the potential field due to impact be first written generally as $U(t)$,
corresponding to a beam that is suddenly switched on or modulated in some
way, perhaps as a sinusoidal potential Us in ωt.

Because we are concerned with a strong field situation modified wave
functions may be expressed as:

$$\Psi^1_n = \sum_{j=0}^{\infty} a_{nj}\psi_j$$

where $\{a_{nj}\}$ becomes the unit matrix as $U \to 0$. Corresponding to a given ω,
providing U is sufficiently large a pair of states Ψ^1_0 and Ψ^1_n should occur in
equal abundance, where

$$\hbar\omega = E^1_n - E^1_0$$

As $\omega \to 0$, corresponding to the case of an electron beam existing for so
long that time variation may be ignored, clearly only the perturbed ground
state may be occupied. But this ground state must be clearly distinguished
from that of the original molecule, in view of the different environment. The
time of transit of a molecule through a beam of dimension 10^{-2} cm is
about 10^{-7} sec and this on the scale of electronic motion, corresponds to a
very large number of cyclic orbits. Thus it is appropriate to regard U as a
static field.

The perturbed ground state $\Psi^1_0 = \sum_{j=0}^{\infty} a_{0j}\psi_j$ and $|a_{0j}|^2$ measures the abun-
dance of the state ψ_j after the molecule leaves the electron beam. There is
thus a clear distinction between steady states in the sense of radiational applied
fields and those due to a beam of particles.

The whole process of fragmentation involves two stages comprising initial
excitation followed by dissociation. The latter process has been discussed
quantum mechanically by Coulson and Zalewski[8], using a direct approach,
and by Hall and Levine[9,10] using the Green's operator with perturbation
potential essentially the interaction between the separating particles. In prin-

ciple this is an infinite term perturbation treatment, but there will be serious practical limitations and we have often to be content with a truncated series or even the first one or two terms. Equivalently, the S-matrix formulation[11] could be employed.

From a discussion of the time-dependent Schrödinger equation for a case where a discrete level can interact with the continuum, Coulson and Zalewski find that

$$|a_{11}|^2 = e^{-2vt}$$

where $v = \pi \varrho \hbar^2 |V|^2$ and ϱ = density of states in the continuum. Thus the initial wave function

$$e^{-iEt/\hbar} \quad \text{becomes} \quad a_{11}e^{-iEt/\hbar}$$
$$= e^{-vt}e^{-iEt/\hbar}$$
$$= e^{-it/\hbar(E-i\Gamma)}$$

where $\Gamma = v\hbar$.

Thus the energy can be considered as a complex quantity. The legitimacy of the steady state treatment of the initial excitation is linked up with the magnitude of Γ. If the half life for dissociation is very short then switching on the electron beam at $t = 0$ with a well-defined energy does not suffice to fill up the line-shape[12,13]. If the duration of the impact is very long compared with the decay time then the processes of impact and dissociation cannot be separately identified. Because energy and time are conjugate variables, uncertainty in either suffices to fill up the line shape. The spread in electron beam energy is sufficient in practice to do this even if impact starts merely at $t = 0$ and is of short duration. Otherwise a beam present in the infinite past ($t = -\infty$) would lead to the same result.

It might be thought that the line shape for metastable processes should be less broad than for fast reactions in the ion-source. Other things being equal this would follow from the nature of the complex energy which has a relatively small imaginary component when the dissociation to continuum states with outgoing waves has a long half-life (of order 10^{-5} sec). The imaginary component $\Gamma = \frac{1}{2}\hbar k$ is small for metastable ions with $k \sim 10^5$ sec^{-1}. Since accessible electronic states are in rapid reversible equilibrium, not only the ground electronic state but others differing not too much in energy can be involved in the mechanism of dissociation. The line breadth has a bearing on the capacity for such states to interact and establish equilibrium. Mass spectral rate processes can be expressed empirically by an exponential, (i.e. Arrhenius) rate factor and an "effective temperature" dependent on emission voltage, source temperature, and the reactive state of the ion. The latter is, in principle, calculable in terms of Green's operator, essentially an infinite perturbation series treatment. To the extent that an empirical effective temperature can be meaningfully assigned, one could expect the canonical density matrix to be relevant to the discussion of such systems. This determines the incidence of various low-lying electronic states in the dissociation mechanism; fewer will be involved for rate processes with a long half life as in metastable dissociations.

Much of the subsequent discussion relates to some anomalous effects observed in metastable ions. A study of such effects might be expected to lead to a better understanding of mechanism of dissociation in general.

3   SOME EXAMPLES OF ANOMALOUS METASTABLES

Metastable ions sometimes appear anomalous either in appearance potential, in effective mass, or in the nature of their pressure dependence. An example of an appearance potential anomaly is shown by the release of nitric oxide from nitrobenzene; whilst our interpretation of this anomaly was at the outset based on a possible pair production mechanism, the difficulty was resolved by a careful application of high resolution techniques. This succeeded in demonstrating clearly the involvement of a wall reaction leading to the production of aniline[14]. (Compare also Budzikiewicz[15].) Mass anomalies seem to be far less common and when they do arise their detailed interpretation can provide some challenging problems. An early example is provided by the mass spectrum of acetylene, as reported by Melton[16] for $C_2H^+ \rightarrow C_2^+ + H\cdot$. We have recently confirmed that this is indeed a genuine pressure independent process and yet it gives a peak at mass 23.10 differing significantly from the normal m_2^2/m_1 value of 23.04.

In a recent short note[17] we have reported a similar striking anomaly in the methanol mass spectrum. Corresponding to the dissociation $32^+ \rightarrow 31^+ + H$ an anomalous peak appears at mass 30.09 differing significantly from the Hipple formula prediction of 30.03 (Figure 1). Possible explanations dependent upon gas phase collisions have been partially successful in predicting an anomalous mass in this region but are not convincing because they would obviously imply a bimolecular process requiring metastable abundance to vary as the square of the pressure in the ion source. Correspondingly, the relative abundance ought to show a linear increase with pressure.

A more plausible interpretation might be that the collision takes place at a solid wall and consequently does not exhibit pressure dependence in view of the invariable surface area thus presented. From the appearance potential studies for nitrobenzene no doubt can remain that wall reactions can and do occur so it would not be so remarkable if mass deviations should not also emerge in appropriate cases. An electron-transfer mechanism has already

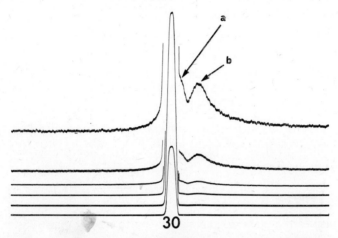

Figure 1 The mass spectrum of methanol in the region of mass 30
a normal peak at mass 30.03   *b* anomalous peak at mass 30.09

been postulated by Newton and co-workers for somewhat similar mass shift anomalies provided by rare gas ions[18]. But their mechanism is basically quite different to the one we are now going to propose, in that it invokes multiple ionization.

We shall suppose that the parent ion (m_1^+), which is involved in the production of a metastable ion in the field-free region consists essentially of an ion-atom complex to be described as $(p-1)^+$ and H· moving together with a common velocity u and having masses $m_2 - \varepsilon$, m_3 where m_2 is the mass of the $(p-1)$ radical, and ε that of an electron, $m_1 = m_2 + m_3$ that of the neutral molecule.

By a resonance mechanism occurring at the monitor slit it may be possible for a neutral molecule of methanol to exchange an electron with the above complex—this transfer can occur in both directions resulting in an ion-atom complex where the $(p-1)^+$ component has been retarded by momentum exchange transactions.

Thus
$$m_1^+ \rightarrow m_2^+ + m_3$$

$$(m_2 - \varepsilon)\, u + m_3 u = m_2 v_2 + m_3 v_3$$

$$(m_2 - \varepsilon)\, u^2 + m_3 u^2 = m_2 v_2^2 + m_3 v_3^2$$

where v is the retarded velocity of the $(p - 1)$ ion after neutralization on picking up an electron and v_3 is the hydrogen atom velocity.

It then follows directly that

$$\left(\frac{v_2}{u}\right)^2 - 2\,(1 - \varepsilon/m_1)\left(\frac{v_2}{u}\right) + (1 - \varepsilon/m_1)\,(1 - \varepsilon/m_2) = 0$$

Since ε/m_1 is almost negligible because it involves a mass ratio of an electron and of the parent molecule one can write

$$v_2/u = 1 + \sqrt{\frac{m_3 \varepsilon}{m_1 m_2}}$$

and

$$m^* = \frac{m_2^2}{m_1}\left[1 + 2\sqrt{\frac{m_3 \varepsilon}{m_1 m_1}}\right]$$

This measures the mass discrepancy owing to momentum transfer by the colliding electrons.

Remarkably enough the predicted deviation from the normal metastable mass for methanol is about right in order of magnitude if we substitute the known masses for m_1, m_2 and ε. For example, with $m_1 = 32$, $m_2 = 31$, $\varepsilon/m_3 = \frac{1}{1837}$, m^* turns out to be 30.085.

But an entirely different mechanism involving hydrogen transfer can also lead to a mass dependence closely similar to the above and in more satisfactory accordance with isotopic substitution effects.

A mechanism involving a "sticky" collision with a thermal hydrogen or deuterium atom leads to the correct mass relationship but fails to explain why the excess energy of impact is not immediately released again as translational energy. This would involve accumulation of excess vibrational energy in the products without peak broadening due to release of kinetic energy. In view of the large internal energies involved this seems quite unrealistic.

For head-on collisions and a forward "glory" effect:

$$p^+ + H\cdot \rightarrow (p + 1)^+ \rightarrow p^+ + H\cdot$$

where the velocities of p^+ and $(p+1)^+$ are respectively u_1 and u_2.

By conservation of momentum

$$m_1 u_1 = (m_1 + 1) u_2$$

and $\frac{1}{2} m_1 u_1^2 = eV$ where eV is the energy of p^+ acquired in the accelerator field.

Thus the translational energy of $(p + 1)^+$ formed in the "sticky" collision

$$= \frac{1}{2} (m_1 + 1) u_2^2 = \left(\frac{m_1}{m_1 + 1} \right) eV$$

but a considerable increment of energy $eV/(m_1 + 1)$ must be accommodated in internal degrees of freedom.

The apparent mass

$$m^* = \frac{m_1^2}{m_1 + 1} \times \text{(energy factor)}$$

$$= \frac{m_1^2}{m_1 + 1} \cdot \frac{m_1}{m_1 + 1} = \frac{m_1^3}{(m_1 + 1)^2}$$

In the absence of a third body to convey away the excess energy it is clearly unrealistic for energy

$$\frac{eV}{m_1 + 1}$$

many times greater than the dissociation energy to be involved as energy of excitation.

In order to ensure that momentum p is unaffected whilst energy is removed we require that $(\Delta p)^2/2m$ is large but Δp, the momentum change, is small, and this condition can only be fulfilled by a light particle such as an electron. Thus the only available suitable third body is an electron but its emission would result in a doubly charged ion and such an ion is known not to be involved.

Even more elaborate mechanisms involving collision with high energy hydrogen can be visualized. For example, an ion of mass 31 produced in the source will acquire a slightly higher velocity in the accelerator field than the parent ion. It is thus possible for hydrogen released from such an ion to give the parent ion a "nudge" forward and thus increase the apparent mass of a collision induced metastable. By simple momentum and energy considerations

$$m^* = \frac{(m_1 - 1)^2}{m_1} \times \phi^2$$

where

$$\phi = 1 + \dfrac{2\,\dfrac{m_3}{m_1}\left(\sqrt{\dfrac{m_1}{m_2}} - 1\right)}{(1 + m_3/m_1)} \approx 1 + \dfrac{m_3^2}{m_1^2}$$

where

$$m_1^+ \rightarrow m_2^+ + m_3$$

For the process $32^+ \rightarrow 31^+ + H\cdot$ the apparent mass should be shifted by a factor 1.00194.

The observed normal metastable has mass $= 30.03$ and the anomalous value $= 30.09$, increased by a factor $= 1.00199$ which is in reasonable agreement.

In the perdeuterated compound CD_3OD
$\phi^2 = 1.00309$ and $m^* = 32.235$ with an energy excess of

$$\tfrac{1}{2}m_1 u_1^2 \left[1 + (m_3/m_2) - (1 + (m_3/m_1))\,\phi^2\right]$$

This latter is 0.352 eV for an 8 kV ion energy, a more plausible value than that associated with a thermal atom. Again the predicted mass is in good agreement with the observed 32.225 and the normal mass position of 32.11. Similar prediction of the anomalous mass for other isomers CD_3OH, CH_3OD can be successfully achieved on this basis.

A possible interpretation of the unexpectedly high collision probability and quasi-unimolecular property may be based on the idea that of the two possible tautomeric forms of the parent ion, one is readily adsorbed on the walls of the ion source.

Thus we have the tautomeric ions

$$CH_3\!-\!{}^+\!O\!\diagup^H \quad \text{and} \quad \cdot CH_2\!-\!\overset{+}{O}\!\diagdown_H^{H\cdot}$$

and ions of the first type may be adsorbed in the form of a monolayer. Their effective concentration is consequently pressure independent and they may be supposed capable of combining with tautomeric ions of the second type to form a doubly charged species:

$$CH_3\!-\!{}^+\!O\!\diagdown^H_{CH_2\diagdown\,{}^+O\diagdown_H^H}$$

which is indistinguishable in m/e from that of the parent ion itself.

This double charged ion may, however, dissociate again in the vapour phase owing to the tendency for double bonding and loss of a hydrogen atom.

$$CH_3 - {}^+O \diagup^H \rightarrow CH_2 = \overset{+}{O} \diagup^H + H \cdot$$

This mechanism consequently provides the source of a contiguous pair of ions having a velocity differential in the accelerator field. The high cross-section for knock-on collision is a property of an ion pair moving along the same or contiguous trajectories. It is also reasonable in view of the Thomson postulate[19] that a neutral species colliding with an ion should retrace its path in the reverse direction after impact. This enables us to consider impact in one dimension and is consistent with the relative sharpness of the anomalous peak, which would be much more diffuse if scattering occurred randomly at all angles.

Thus a variety of mechanistic interpretations of mass shift is possible but there are detailed objections to all of them.

4 ELIMINATION OF CO FROM NITRONAPHTHALENES

High resolution techniques and the study of metastable peaks have thrown considerable light on the mechanism for release of neutral CO (and NO) from aromatic nitro compounds. It has been suggested[20] that the release of CO from nitronaphthalenes (metastable peak at $121.5u$) can be correlated with electron deficiency at the 8-position. Evidence for this has been based on detailed m.o. calculations. It may however be worth mentioning that a useful interpretation can be based on simple properties of the least bonding orbital. A striking result was the anomalously low release of CO from 1:3 dinitro-naphthalene, in view of the favourably low electron deficiency in the 8-position.

It has been remarked[21] that the electron distribution due to the nitro-group in nitrobenzene is qualitatively similar to that of the benzyl cation. This is in accordance with the formal positive charge on the nitrogen atom in NO_2. It is reasonable to regard this as a general property of the nitro group as a substituent in aromatic ring compounds.

Corresponding to the single N.B.M.O. in the benzyl radical there exists a pair of such N.B.M.O. in

On a naive m.o. basis we may discuss electron deficiencies in the corresponding cation as though directly determined by this pair of non-bonding orbitals.

Written in the form with all the electron deficiency located on one nitro group it is clear that the m.o. are not strictly orthogonal.

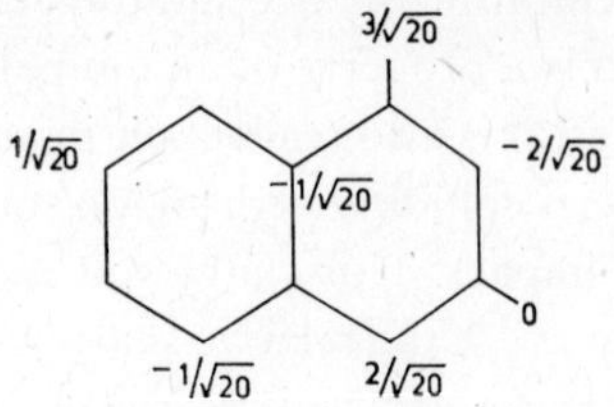

(A)

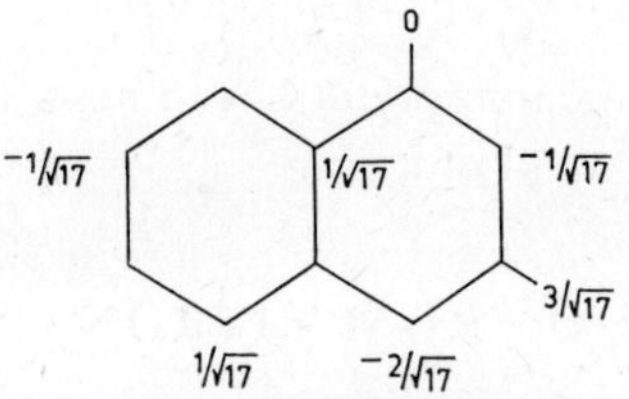

(B)

Although (A) meets all the requirements of zero electron deficiency in the 8 position and maximum electron deficiency on the substituent yet it cannot correspond to a true stationary state because it is not properly orthogonal to (B). Although it is possible to imagine that (A) can exist as a time-dependent state during the actual process of specific rearrangement and accompanying hydrogen transfer, yet this would imply that vibrational quanta were unfavourably restricted in order to provide the extra electronic energy. Such a mechanism would therefore be statistically less favourable.

It is therefore reasonable to suppose that the requirement of maximum energy in the vibrational modes leads to that combination of A and B leading to strict orthogonality. The obvious choice is $(A + B)/2$ and $(A - B)/2$. Quite clearly, however, this can only be achieved at the sacrifice of considerable electron deficiency in the nitro group for we have now, substantially, a mean value of the more asymmetrical distributions in A and B, i.e.

$$\frac{1}{2} \times \frac{9}{20}$$

$$\frac{1}{2} \times \frac{9}{17}$$

on the respective groups.

This lowered electron deficiency is apparently insufficient to facilitate release of CO. In 1 nitronaphthalene the electron deficiency is distributed as in (A) with $\frac{9}{20}$ of an electronic charge located on the nitrogen. This together with favourable electron deficiency in the 8-position presumably accounts for the strong tendency to release CO as $C^- \equiv O^+$ (Figure 2).

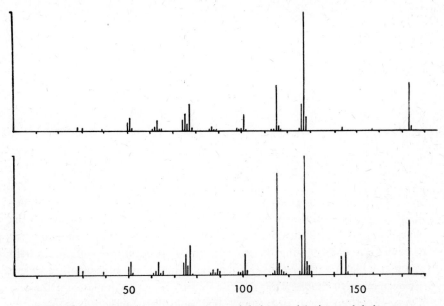

Figure 2   The mass spectra of 2-nitronaphthalene and 1-nitronaphthalene

Substitution of a second nitro group in either the 4 or 5 position produces a splitting of the energy levels on account of non-zero coefficients at these atoms in the monosubstituted case. The reason for the reluctance of the 1:5 isomer to release CO is apparent from a comparison of charge densities in the corresponding naphthoquinodimethane. According to the *Dictionary of Values of Molecular Constants*[22] the following distributions apply to the highest energy occupied orbitals. The striking differences in electron deficiencies in the 8-position is at once apparent.

More refined calculations using a self-consistent field method[23] have confirmed very limited electron transfer between the nitro group and the benzene ring. The migration away from the ring into the nitro group appears as small as 0.016 of an electronic charge. It would appear therefore that the best calculations on charge distribution in the neutral molecule do not accord well

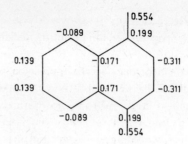

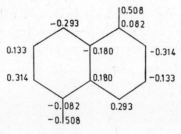

with the suggested fragmentation mechanism. From the qualitative stand-point it may be said that the cations undergo fragmentation as though the charge were delocalized. One reason for this is probably that the tendency to remove a non-bonding oxygen electron produces an electrostatic field effect in which charge repulsions can be reduced by delocalization of the nitrogen positive charge. Thus the distribution is better simulated by the odd alternant cation than it is in the neutral molecule. Interaction between the

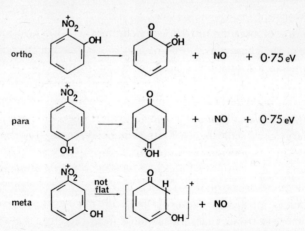

Figure 3 Release of nitric oxide from isomeric nitrophenols

non-bonding oxygen charge and the Π-electron system (a species of $\sigma-\Pi$ or $n-\Pi$ interaction) is apparently an essential feature of the aromatic cation.

Similar studies of metastable peak widths permit an evaluation of resonance integrals where effects can be attributed to Π-electrons. For example, the remarkable result that o- and p-nitrophenol both release NO with the accompaniment of 0.75 eV kinetic energy whilst the meta isomer exhibits no peak broadening (Figure 3) would appear to be a property of alternant ring systems. Assuming that the isomeric transition states have the same energy, the energy releases for ortho and para isomers can be attributed to the greater stabilities of the product ions. A similar simplified approach to that in the discussion of nitro-naphthalenes allows the additional resonance energy to be estimated.

Consider the following reaction sequence:

The particular formulation of the transition state C is in accordance with the steric repulsion due to the substituents simultaneous bonded in the para position. Hydrogen transfer from the adjacent carbon is considered to involve the remote oxygen cation which becomes trivalent. Whilst hydrogen transfer to the adjacent oxygen is possible this seems unlikely since it would lead to an increase rather than a decrease of steric repulsion. It should now be remarked that the original conjugated system involving 10 atoms in B is reduced to a pentadienyl radical in C. Moreover the energy of the latter is clearly the same for all isomeric forms. On the other hand the product ion D is clearly much more stable than the meta isomer which can be formulated only with 3 double bonds viz:

The additional resonance energy in the *o*- and *p*-isomers is $\sim 2\beta_{co}/\sqrt{7}$ = 0.75 eV whence

$$\beta_{CO} = 95.8 \text{ kJ/mole} \quad (22.9 \text{ kcal/mole}).$$

This compares satisfactorily in order of magnitude with the commonly adopted value for thermochemical applications ($\beta_{CO} = \sqrt{2}\beta_{CC} = 24.0$ kcal/mole) and lends credibility to the Π-electron interpretation. It should, however, be noted that this interpretation fails completely if the tautomeric form of the transition state C without hydrogen transfer is invoked. For in that case the transition states will differ in energy because the divalent oxygen cation is capable of double bonding to the pentadienyl radical. This should imply that the *o*- and *p*-isomers would have relatively stable activated states and the *m*-isomer would contain excess electronic energy. Since this is contrary to the evidence from metastable peak widths, it seems reasonable to conclude that the activated states are indeed subject to hydrogen transfer. This illustrates a further possible applications of the metastable technique in allowing conclusions to be drawn on the structures of activated states which could not be conceivably made in any other way.

McLafferty[24], Williams[25], Mandelbaum[26] and others have referred to the importance of the positive charge localization requirement and its dependence on the nature of the substituent groups. For example, it has been shown that elimination of the methoxy radical from

$$\text{MeO}-\langle\ \rangle-(CH_2)_3\ C\underset{OMe}{\overset{O}{<}}$$

is much smaller than from the nitro substituted form

$$\text{NO}_2-\langle\ \rangle-(CH_2)_3\ C\underset{OMe}{\overset{O}{<}}$$

an effect attributed to the presence of positive charge on the ring in the first case. The rate of fragmentation is believed to be strongly correlated with the location of positive charge on the keto moiety.

In the substituted dinitronaphthalene it is observed that replacement of one of the two nitro groups by an amino group does not inhibit completely the CO release mechanism, which (cf. butyrates) might be imagined to result from a change in location of the positive charge. The methoxy and amino substituents should both exhibit electron donor ($-E$) properties in view of the pairs of p_z electrons able to conjugate with the ring.

It seems most unlikely that the ionization is entirely localized in the NO_2 group or entirely due to a Π-electron ionization. If it were localized the major differences between, for example, 1:4 and 1:5 derivatives would be difficult to account for in view of the S.C.F. results and negligible interaction with the ring. Moreover, the assumption of complete delocalization would imply approximately one quarter of the positive charge available at each nitro group in the dinitro compounds. Since reaction rates are exponentially dependent on activation energies, which are themselves varying with positive charge densities, it is inconceivable that a 50% change in positive charge would fail to exert a pronounced effect on the release of CO.

The choice therefore remains between a dynamic transfer of electron excitation between Π and n orbitals or in some stationary state of the ion that involves differing configurations. Whilst an even electron system such as that of the neutral molecule can be represented well by a single determinantal wave function the open shell character of the ion will allow differing configurations to be formulated corresponding to determinants with either one half-filled Π or half-filled σ orbital. The contribution of the former will provide a greater sensitivity to substituents than is the case in the closed shell system.

This open shell character and interlacing of σ and Π energy levels provides one of the reasons for the ambiguity in the description of ionic species in many mass spectral mechanismus. Wherever the σ and Π separation is not preserved, geometrical properties such as the existence of a plane of symmetry become correspondingly less well defined (and conversely). However such changes will most likely arise as the amount of excitation energy entering the ion increases. Those ions that have less electronic excitation will be statistically predominant because they permit a greater amount of excitation to occur as vibrational quanta, with a consequent enlargement of phase-space volume.

In those cases where the quantum numbers corresponding to additional geometrical symmetry elements (such as a plane of symmetry in aromatic systems) no longer retain sharp values it is evident that a statistical treatment of the valence electrons becomes more appropriate. Thus highly activated positive ions that undergo extensive rearrangement will no doubt be amenable to a semi-classical statistical discussion of charge distribution. An intermediate stage is one where configuration interaction is extensive and intramolecular repulsion between charged atoms may be assessed in the light of the Hellmann–Feynman theorem.

JAHN-TELLER EFFECTS IN HIGHLY SYMMETRICAL 3-DIMENSIONAL IONS

The benzene cation contains a degenerate pair of orbitals of e_{1g} symmetry type. The ground state of benzene may be designated $a_{1u}^2 e_{1g}^4$ but after loss of an electron to form a positive ion $(a_{1u}^2 e_{1g}^3)$ either of the two available e_{1g} orbitals can be singly occupied. This leads to a distortion of the nuclear framework in which the cation can flop continuously from one to another of three possible geometrical configurations. In accordance with the Jahn–Teller principle[27], such an ion will readily distort from the initial fully symmetrical form in which it is instantaneously produced by electron impact. For a similar reason CH_4^+ is believed to be exactly or nearly planar. This effect is consequently not merely a second order perturbation but one involving vibrational modes of very large amplitude; it is therefore a type of direct interaction between electronic and vibrational energy capable of producing severe instability. It might be expected that 3-dimensional highly symmetrical ions such as that of hexamethylenetetramine would exhibit similar gross distortional behaviour and this seems consistent with the unusual mass spectra.

The positive ion of hexamethylenetetramine is quite remarkable in that it appears to undergo fragmentation into subsystems of small mass, neither the parent ion nor even the $(p-1)^+$ exhibiting particular stability. In the latter respect it differs markedly from some other cyclic amines such as pyrrolidine and piperidine, where the loss of a single hydrogen stabilizes the ion and provides the base peak in the mass spectrum. (For a discussion of metastable transitions vide Saunders and Williams.[28]) This is perhaps surprising in view of the apparently rigid 3-dimensional structure of the parent molecule. Substantially, hexamethylenetetramine is comparable structurally to P_4 but with 6 methylene bridges between the apical nitrogens. However, the strain cannot be attributed to strain in the bond orbitals as in the P_4 case because the methylene bridges permit the bonding atomic orbitals to take up a natural inclination.

In the case of hexamethylenetetramine it is evident that electron impact will lead to loss of a non-bonding electron from any one of four equivalent nitrogens. In the T_d group for tetrahedral symmetry we can single out the 8 initially available non-bonding electrons and distribute them amongst the totally symmetrical representation A_1 and the triply degenerate one T_2, thus $A_1^2 T_2^6$. In the cation, the configuration $A_1^2 T_2^5$ of the non-bonding electrons is triply degenerate and the ion will show a tendency to flop between 4 shapes in which one or other of the 4 vertices is displaced from the other three. The resulting gross distortions lead to fragmentation into small subsystems in which the following ion

$$CH_2 \overset{\displaystyle N^+}{\triangle} CH_2$$

is prominent.

It is interesting to notice that there is a "diatomic" analogue of the above system, namely diazabicyclooctane. Although a 3-dimensional molecule from the point of view of the sub-system of electrons that are involved directly in the process of ionization it is quasi diatomic, in that a single pair of nitrogens with $-CH_2-CH_2-$ bridges is involved. In accordance with the Jahn–Teller principle distortion should not occur upon ionization. The relevant symmetry group if we think only of N—N is $D_{\infty h}$ and the 4 non-bonding electrons are assigned to the configuration $A_{1g}^2 A_{1u}^2$.

On ionization the non-bonding electrons do not form a degenerate system, since A_{1u} is not a multiple representation. This is in line with the less specific fragmentation along a variety of energy surfaces. Whilst the base peak still appears at mass 42, an additional ion $C_3H_5N^+$ is also prominent which indicates the importance of bond cleavage without rearrangement of methylene groups. For if rearrangement occurred in hexamethylenetetramine and the predominance of $C_3H_5N^+$ in diazabicyclooctane were due to thermodynamic stability of the latter it would be expected to show up also in the former. Actually it is quite insignificant.

It may be remarked that the parent ions produced instantaneously on impact should have a completely delocalized positive charge. The Jahn–Teller effect leading to charge localization is a relatively slow one depending upon the nuclear distortions. Initially it is energetically more reasonable to place one quarter of an electron on each equivalent nitrogen. One has only to conceive of a process in which 3 of the fractional charges are removed and placed together with a fourth one. The energy required to remove them is $\frac{3}{4}I$ where I is the ionization potential; whilst this is regained on placing them back on a single nitrogen there will be a greater repulsion between them when brought close together. If it is correct to suppose that the p^+ ion starts off in a configuration of minimum energy then the smoothed out positive charge will be a necessary requirement.

It is still true that the smoothed-out charge is less repulsive in diazabicyclooctane than in H.M.T. For in the latter we have $^4C_2$ repulsions giving a total electrostatic repulsion* of $6e^2/16r$ and in the former the repulsion is $e^2/4r^1$.

Thus electrostatic repulsion effect

$$\frac{\text{H.M.T.}}{\text{D.B.O.}} = \frac{3}{2}\left(\frac{r^1}{r}\right)$$

where
$$r = 2.40 \text{ Å}, \quad r^1 = 2.51 \text{ Å}$$

These electrostatic energies are large in comparison with the excess energies usually considered sufficient to provide rapid fragmentation in the ion source. This is considered less than 1 eV by Williams *et al.*[29], which may be compared with the respective values 1.50 eV and 1.43 eV for Coulomb repulsion in these cases.

The more important peaks in the mass spectra are shown below.

Table 1   Important ions in mass spectra

H.M.T.		Diazabicyclo-octane	
140	25	112	52
42	100	70	16
28	36	58	87
		57	39
		56	37
		55	82
		53	80
		42	100
		41	30
		30	16
		29	19
		28	10

The detailed quantum mechanical evaluation of a 3-dimensional poly-atomic cation such as hexamethylene tetramine would present a problem of unmanageable complexity, involving no less than 55 valence electrons. Even

---

* Although the concept of direct electrostatic interaction between fractional charges presents no difficulties for an even electron cation, the situation is different for an odd number of electrons owing to the absence of the corresponding 2-electron Coulomb repulsion integral. But the change in Coulomb parameter $\delta x = -e^2/R$ when unit positive charge is situated at distance R from a particular atom leads to an effect substantially equivalent to that of direct electrostatic repulsion. This is inherently reasonable since charges, irrespective of their origin should be capable of interaction.

for the simplest triatomic molecular systems the calculated molecular conformations—of crucial importance to fragmentation behaviour—depend to a high degree on values for multi-centre integrals. For the more important 2-centre integrals it is often supposed[21] that

$$H\mu\nu = 0.5K \, (H\mu\mu + H\nu\nu) \, S\mu\nu$$

but particular choices of K can lead to wide variation in the calculated bond-angles. For 3-dimensional molecules and molecular ions of more critical stability it would clearly be unrealistic to expect to reach useful conclusions. This suggests that to achieve some insight into the structural properties of large polyatomic ions a semi-classical electrostatic model may be more appropriate.

6 SEMI-CLASSICAL MODELS FOR HIGHLY SYMMETRICAL CATIONS

In this model[30] the electron densities and potentials are required to satisfy Poisson's equation:

$$\nabla^2\phi = -4\Pi ne$$

and the electron density

$$n = 8\Pi/3h^3 \, (2me)^{3/2} \, [a + (V - V_0 + a^2)^{1/2}]^3$$

where $a = (2me^3)^{1/2}/h$ if exchange effects are included, otherwise zero.

As we have already remarked the lower state of the ion $(CH_2)_6N_4$ is one where there is an equal probability of each of the nitrogens having lost an electron from a non-bonding orbital. The resulting positive charge may be considered to be smoothed out and distributed uniformly over a sphere passing through the vertices of the tetrahedron. A total of 20 units of positive charge from the nitrogen atom nuclei and K shell electrons may likewise be distributed.

The smoothed out density of positive charge is obviously:

$$\sigma = |e|/4\Pi R^2$$

At the sphere of radius R, two considerations must be satisfied:

$$(V_1)_R = (V_2)_R$$

$$-\left(\frac{dV_1}{dr}\right)_R + \left(\frac{dV_2}{dr}\right)_R = 4\Pi\sigma$$

An important feature of the semi-classical treatment of electron distribution is that it permits of a general formulation of non-stationary states or a prepared ion in a mixture of states through use of the Bloch equation[31]:

$$\frac{\partial C}{\partial \beta} = -HC$$

and the so-called canonical density matrix

$$C(\mathbf{r}, \mathbf{r}_0, \beta) = \sum_i \psi_i^*(\mathbf{r}) \, \psi_i(\mathbf{r}_0) \exp(-E_i\beta)$$

where $C(\mathbf{r}, \mathbf{r}_0, 0) = \sum_i \psi_i^*(\mathbf{r}) \, \psi_i(\mathbf{r}_0) = \delta(\mathbf{r} - \mathbf{r}_0)$ is the infinite temp. value ($\beta = 1/kT$).

Whilst the treatment more particularly applies to many-body problems as in metals, it seems likely that certain large isolated ions in excited states can be well described by a statistical model, bearing in mind, however, that the exponential factor appropriate to a real temperature might be more appropriately replaced by a weighting factor $[(E - E_0)/E]^{n-1}$ for the case of n effective vibrational oscillators. But the usual Q.E.T. assumption of sufficient low lying electronic-states to form a continuum is just the requirement in the Theory of Metals for a discussion of energy bands and density matrices.

Those ions dissociating instantaneously in the ion-source will be characterised by a temperature such that the canonical density matrix approaches the $C(\mathbf{r}, \mathbf{r}_0, 0)$ value. Metastable ions will correspond to a much lower temperature and amongst the infinite number of terms in the density matrix those near the beginning of the series will be much more important because of the weighting factor.

An interesting feature of the semi-classical approach is the close analogy to other well-known physical phenomena. For example, it may be shown that a solution of the Bloch equation for an especially simple case of an electron gas without interactions is the Gaussian form[32]

$$C(\mathbf{r}, \mathbf{r}_0, \beta) = \exp(-R^2/2\beta)/(2\Pi\beta)^{3/2}$$

where $R^2 = (\mathbf{r} - \mathbf{r}_0)^2$, a not altogether unexpected result in view of the similarity of the heat conduction equation[33]:

$$\nabla^2 \theta = \varkappa \frac{\partial \theta}{\partial t}$$

and the Bloch equation for a free electron:

$$\nabla^2 C = - \frac{\partial C}{\partial \beta}$$

This suggests that the electron distributions in ions that largely determine the fragmentation patterns should be formally related to well-known results for heat transfer in systems of similar geometry and equivalent boundary conditions.

Due recognition of the mathematical similarities between certain well-known theories in classical physics and the corresponding ones in wave mechanics might thus provide valuable insight into fragmentation phenomena in general. In addition to the above correspondence between density matrices for the type of many-body problems arising in polyatomic organic ions and heat transfer properties, one might also mention such notable instances as the Mie theory of light scattering[34] on the one hand and the corresponding wave-mechanical treatment of electron impact.

7 RELEASE OF ENERGY FROM AROMATIC RING SYSTEMS

It would be instructive to seek examples of possible correlations between thermochemical properties such as forward and reverse activation energies and electron distributions in metastable ions. Spiteller[35], Santoro and Spadaccini[36], Hirota[37] and others have emphasized that any correlation with electron populations in the ground state could in principle only be detected for very low energy impact processes, otherwise excited states are implicated. We have shown that cross-sections for ionization of polycyclic hydrocarbons under normal electron impact conditions are in accordance with an almost equal facility to release any available Π-electron, irrespective of its energy level in the molecule[38]. Moreover, certain processes such as the loss of a methyl radical from isomeric alkanes, do not appear to depend on electron density in the particular C–C bond but rather on the bond strength of the neutral parent compound. This again is in accordance with excitation of a band of energy levels so that the observed fragmentations suggest a smoothed out electron distribution resembling the parent compound rather than one particular state of the parent ion. The use of an effective "temperature" to describe the actual degree of internal excitation is in accordance with the view that the randomization implied by QET means that the energy is distributed in the same way "as though the parent ion had a fixed temperature"[39]. In a similar context Dougherty[40], in putting forward a perturbation molecular orbital treatment of positive ions, has claimed that the initial states of a molecule ion can be approximated by a set of doublet states that correspond to removal of a valence electron from any of the levels of the parent molecule. From the successful application of Hammett's relation he also infers that $k = A\exp(-E^*/E)$ where $E =$ internal energy, $E^* =$ activation energy. With $T =$ "Effective Temperature", $E = RT$ (compare also [41]).

An important feature of the metastable ions is that, irrespective of the electron impact energy, they depend on states of the ion that have only slightly more energy than the threshold value for dissociation. Thus we can expect to detect effects that are simply related to the ground state without the complication of electronic excitation appearing in the mass spectrum as a whole. In other words the mass spectrometer is acting as an energy filter for those ions having conveniently simple properties and these may be profitably studied apart from the gross behaviour. The latter is dependent upon ions in prepared states representing some statistical average over low-lying energy levels (β, the density matrix parameter, comparatively small).

An example that comes to mind concerns the release of neutral isoconjugate species from a pair of related dissociations. Ottinger[42] has shown that the release of acetylene from benzene is accompanied by a release of

Table 2   Fragmentation energies and corresponding ion lifetimes for the metastable decomposition $C_4H_4N_2^{+\cdot} \rightarrow C_3H_3N^{+\cdot} + HCN$ in pyrazine

Ion accelerating voltage (kV)	T (mean at $\frac{1}{2}$ height) (eV)	Source residence time (t_1) (μsec)	Time of flight (t_2) (μsec)	Lifetime of dissociating ion ($t_1 + t_2$)
8	0.02221	1.4	8.1	9.5
4	0.02106	1.9	11.5	13.4
2	0.01969	2.5	16.1	18.6
1	0.01872	3.4	22.9	26.3

0.031 eV $\pm$ 0.005, and with a metastable intensity of 0.0036. For the corresponding release of HCN from pyrazine we have measured an energy release of 0.022 eV with an ion accelerating voltage of 8 kV. Ottinger used a much lower accelerator field (≈ 200 V/cm). In the pyrazine measurements the effect of metastable ion lifetime on energy release was evaluated (Table 2).

S.C.F. calculations on azines[43] and charge densities indicate values as follows:

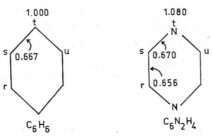

To localize a pair of Π electrons in a C–C or C–N bond requires energy

$$2\,[\beta_{rs}p_{rs} + \beta_{st}p_{st} + \beta_{tu}p_{tu}]$$

where β_{ij}, p_{ij} are respectively resonance integral and bond order.

The resonance integrals β^{cc}, β^{CN} are closely similar. Thus the localization energies are respectively 4β and 3.992β where β is the common value. There is nothing here to indicate any significant difference in activation energy in the forward direction. What we observe as a release of K.E. concerns a reverse activation energy and if we may attach any significance to the somewhat lower value for pyrazine, this might perhaps be attributed to the fact that a triple $C\equiv N$ bond is 9 kcal/mole weaker than a $C\equiv C$.

8   DISTRIBUTION OF EXCESS ENERGY IN THE FRAGMENTED IONS

It is usually assumed that excess energy involved in forming the activated state is released substantially or entirely as kinetic energy. Thus, in forming the final product species there is involved a decrease in total energy relative to the activated state; the above assumption would therefore imply equality of kinetic energy release with the reverse activation energy. It is, of course, true (and we discuss this later) that somewhat more energy than the bare minimum is required in order that dissociation may proceed at the requisite rate for detection as a meta-stable. Thus some increment of excess internal energy is observed as excess translational energy. But this is a refinement leading to peak shape variability. Properly evaluated, it can lead to additional information on the number of effective oscillators in the metastable ion. The immediate question concerns the validity or otherwise of the broad assumption that the reverse activation energy and kinetic energy release (as measured by peak width) may be equated. This does not imply that there are no vibrational quanta left in the products (as is well known these are often sufficient to produce a secondary fragmentation) but that they may be equated with those in the reactant species and consequently do not affect the energy balance. Physically, this might be considered to signify that at a certain critical elongation a bond begins to break apart and thereupon completes the whole subsequent dissociation almost immediately. There is thus insufficient time for additional excitations in other sections of the ion to transfer their excess energy into the dissociation coordinate. Otherwise expressed, we consider that the bond dissociates completely from its ctitical state in about one half of a vibrational period whereas the energy is transferred internally in much longer times of at least several periods. If other bonds reform as in a rearrangement process the principle remains the same.

There are essentially two ways of considering the release of kinetic energy: (i) from the standpoint of statistical mechanics as more appropriate for large polyatomic molecules, (ii) in terms of kinetic models. In (i) we may use statistical criteria such as the higher volume in phase space when energy is released in the kinetic form. In (ii) we may have recourse to related mechanistic considerations for thermally activated processes in ordinary chemistry. Thus it is usually regarded that if A has sufficient thermal energy to undergo reaction when it collides with BC:

$$A + BC \rightarrow AB + C$$

then the selected path through the col of the potential energy surface is one lying at the lowest level passing centrally and symmetrically through the floor of the valley in the energy surface[44]. Transverse oscillations leading to vibrational energy in the products will imply that the reactant species possesses more energy than is otherwise needed to overcome the energy barrier and can be regarded as energetically improbable. Exceptionally, for potential energy surfaces of suitable shape (the so-called "attractive" potential energy surfaces), and notably for such reactions[45] as:

$$H + Cl_2 \rightarrow HCl + Cl$$

$$Cl + Na_2 \rightarrow NaCl + Na$$

the products may be formed with an excess of vibrational energy. This is really tantamount to saying that motion along the reaction co-ordinate may occasionally occur sufficiently energetically for the time scales of transverse and longitudinal motion to become comparable; in such cases interchange of energy becomes possible. Analogous criteria presumably apply to unimolecular dissociations. Clearly then, there cannot be any rigorous demonstration that energy release is rigorously and exclusively kinetic but in many cases, perhaps the most important from the standpoint of metastable ions, it would appear that the assumption is broadly acceptable. This is really implying something about the shapes of the more commonly occurring energy surfaces. It is also empirically reasonable in view of the internal consistency of a variety of results in this field, invoking this assumption.

It would be invaluable if these tentative ideas could be cast into a more definite quantitative form. Notably, Wall, Hiller and Mazur[46] have used an approach that combines features of a classical and quantum mechanical model; in this respect it is exactly in line with the general type of procedure that we have been advocating. Briefly, they have treated particle motions classically and then assumed that the average of several possible motions simulates closely the quantal behaviour. They find that about $\frac{1}{6}$ of the vibrational energy in the reactants can contribute to the energy of activation but the main contribution is translational. What is now required is a comparable model for unimolecular mechanisms that will help to underline the more important features.

Those processes, important for the major metastable ions, generally involve rearrangement of atoms and two or more co-ordinates must appear in the configuration space. For tractability it is generally desirable to envisage a simplified two-dimensional potential energy surface.

Consider the release of acetylene from benzene[14]. We suppose that a vertical plane of symmetry is preserved and consider the 2-dimensional potential energy surface relating to the motions of 3 particles, the two methine groups (A, B) and $C_4H_4^+$ ion (C). We ought to consider internal vibrations in C as well but this provides unmanageable complications.

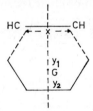

Let G be the centre of mass, M the mass of $C_4H_4^+$ and m the mass of C_2H_2.

The kinetic energy

$$T = \tfrac{1}{2}M\dot{y}_2^2 + \tfrac{1}{2}m\dot{y}_1^2 + \tfrac{1}{2}\mu\dot{x}^2$$

where μ = reduced mass for C_2H_2 i.e.

$$\frac{1}{\mu} = \frac{2}{m} + \frac{2}{m} \quad \text{or} \quad \mu = m/4$$

By conservation of momentum,

$$m\dot{y}_1 = M'\dot{y}_2$$

whence

$$T = \frac{1}{2} m \left(1 + \frac{m}{M} \right) \dot{y}_1^2 + \frac{1}{2} \mu \dot{x}^2$$

The kinetic energy in this model where internal motion of $C_4H_4^+$ is disregarded is diagonalized in the two co-ordinates x and y. Thus the potential energy surface has orthogonal axes.

In that region of the potential energy surface where the system is moving towards the final state ($y_1 \to \infty$) the reaction path is unidirectional and energy interchanges do not occur. The important affects will be associated with the sharp corner where the system approaches from initial $x + y$ ($x = y_1$ = equilibrium C=C bond length) and departs towards a situation where x is the contracted C$\equiv$C bond distance and $y_1 \to \infty$.

The transmission coefficient[7] relating to a change in vibrational quantum number $k \to j$ is determined by the matrix coefficient:

$$Gkj = \frac{(-1)^{k+j} \, 2\pi^2 kj \sin \dfrac{2\Pi p_k l}{h} \, e^{(-2\Pi i/h)p_j l}}{\dfrac{2\Pi}{h} p_j l \left[\left(\dfrac{2\Pi}{h} p_k l \right)^2 - \Pi^2 j^2 \right]}$$

The time to dissociate is of order 10^{-5} sec. This implies that p_k, the momentum along the reaction co-ordinate, is small relative to internal vibrational motion in the sense that

$$\frac{2\Pi}{h} pkl \ll \Pi j \quad \text{for} \quad j > 1.$$

Total energy in the state $\exp\left(\dfrac{2\pi i}{h} p_j x \right) \sin \dfrac{\pi j y}{l}$ is given by:

$$\frac{h^2}{8m} \left[\frac{4p_j^2}{h^2} + j^2/l^2 \right]$$

The zero-point level corresponds to $j = 1$ since $j = 0$ makes the wave function vanish identically. It follows that if the system starts out with $k = 1$, $j = 2$ is incompatible with energy conservation. Thus the sharp corner cannot induce vibrational excitation for comparatively low energy processes involved in metastable dissociation.

Substantially this merely restates the qualitative idea that metastable dissociations proceed with insufficient energy in the reaction co-ordinate (s) to generate vibrational quanta ($j > 1$) in the product species.

In ordinary slow thermal reactions the release of kinetic energy predominates over emission of radiation. Statistically, a single quantum represents a large packet of energy and is a less economical way of distributing energy than in the kinetic form, continuously distributed in phase space. Maximization of phase-space volume is not favoured by large single energy quanta $h\nu$. Whilst reactions in solution provide the necessary degrees of freedom for statistical arguments to be valid, this may not be so for small ions dissociating unimolecularly or for polyatomic molecules of high symmetry where selection rules and forbidden transitions apply. However, these no longer apply when the symmetry of the molecular skeleton is lost in the gross deformations resulting from excitation by electron impact.

If $h\nu$ is involved it must also appear in the reverse sense of activation—this was the original Perrin[47] hypothesis for radiational activation later largely discredited in favour of a collision hypothesis.

9 INTERNAL ENERGIES OF FRAGMENT IONS (HESS' LAW)

This section is concerned with some of the empirical evidence concerning release of kinetic energy and those factors that govern it. Supporting evidence may sometimes be based on parallel reactions involving primary and secondary decomposition processes, as in the stepwise elimination of NO and CO from nitronaphthalene.

$$
\begin{array}{ccc}
& C_{10}H_7O^+ & \\
{}^{-NO}\nearrow & & \searrow{}^{-CO} \\
{}^{m^*=118.2} & & {}^{m^*=92.5} \\
C_{10}H_7NO_2^+ & & C_9H_7^+ \\
{}^{m^*=121.5} & & {}^{m^*=91.2}\nearrow \\
{}^{-CO}\searrow & & \swarrow{}^{-NO} \\
& C_9H_7NO^+ &
\end{array}
$$

Here it is found that the thermochemical consequences of Hess' law appear to be applicable, so that the final states are independent of the sequential details of the reaction mechanism.

In the primary processes involving the molecular ion the metastable peaks are observed to be broad and dish-shaped indicating that the charged and neutral fragments are formed with release of kinetic energy. Those for the secondary decompositions are gaussian in shape but their slopes are significantly different, the slope for elimination of CO being significantly smaller than for NO.

For the broad primary metastables:

$$p^+ \rightarrow (p - NO)^+ + NO^{\cdot} + 0.48 \text{ eV}$$

$$p^+ \rightarrow (p - CO)^+ + CO^{\cdot} + 0.15 \text{ eV}$$

In the case of secondary decompositions:

$$C_{10}H_7O^+ \rightarrow C_9H_7^+ + CO + 0.07 \text{ eV} \quad (\tfrac{2}{3} \text{ height})$$

$$+ 0.14 \text{ eV} \quad (\tfrac{1}{2} \text{ height})$$

$$C_9H_7NO^+ \rightarrow C_9H_7^+ + NO + 0.02 \text{ eV} \quad (\tfrac{2}{3} \text{ height})$$

$$+ 0.04 \text{ eV} \quad (\tfrac{1}{2} \text{ height})$$

Appearance potentials of $C_9H_7^+$ formed by the two possible routes can be determined by operation in the "meta-stable mode", that is by changing the ion accelerating voltage with electrostatic analyser and magnetic sector conditions held fixed. Values for primary decompositions can then be evaluated using a log abundance method and for secondary ones, with slopes different from the molecular ion, by an extrapolated voltage difference method. Only relative values are significant.

Route		A.P. (eV)
A	$C_{10}H_7NO_2^{+\cdot} \rightarrow C_{10}H_7O^+ + NO.$	1.44 ± 0.03 eV
	$C_{10}H_7O^+ \rightarrow C_9H_7^+ + CO$	2.95 ± 0.03
B	$C_{10}H_7NO_2^{+\cdot} \rightarrow C_9H_7NO^{+\cdot} + CO$	1.52 ± 0.03
	$C_9H_7NO^{+\cdot} \rightarrow C_9H_7^+ + NO^{\cdot}$	2.56 ± 0.03

If we designate the parent ion as $[R_1R_2R_3]^+$ where $R_1 = C_9H_7$, $R_2 = CO$, $R_3 = NO$, then:

$$[R_1R_2R_3] + e \rightarrow [R_1R_2R_3]^{+*} + 2e \rightarrow [R_1R_2] + R_3$$
$$+ (T + U) \{[R_1R_2]^+ + [R_3]\}$$

$$[R_1R_2R_3] + e \rightarrow [R_1R_2R_3]^{+*} + 2e$$
$$\downarrow$$
$$[R_1R_2]^+ + T\{[R_1R_2]^+ + [R_3]\} + U\{[R_1R_2]^+ + [R_3]\}$$
$$+ R_3$$
$$[R_1R_2]^+ \rightarrow [R_1]^+ + [R_2] + T\{[R_1]^+ + [R_2] + U\{[R_1]^+ + [R_2]\} \cdots$$
$$- U[R_1R_2^+] \tag{A}$$

Similarly if R_2 is first eliminated:

$$[R_1R_2R_3]^+ + e \rightarrow [R_1R_2R_3]^{+*} + 2e$$
$$\downarrow$$
$$[R_1R_3]^+ + [R_2] + T\{[R_1R_3]^+ + R_2\} + U\{[R_1R_3] + R_2\}$$
$$[R_1R_3]^+ \rightarrow [R_1]^+ + [R_3] + T\{[R_1]^+ + [R_3]\} + U\{[R_1]^+ + [R_3]\} \cdots$$
$$- U[R_1R_3^+] \tag{B}$$

From (A),

$$\text{A.P. } [R_1^+] = \Delta H_f[R_1^+] + \Delta H_f[R_2] + \Delta H_f[R_3] - \Delta H_f[R_1R_2R_3]^+$$
$$+ (T + U)\{[R_1R_2]^+ + [R_1]^+ + [R_2] + [R_3]\} - U[R_1R_2^+]$$

From (B),

$$\text{A.P. } [R_1^+] = \Delta H_f[R_1^+] + \Delta H_f[R_2] + \Delta H_f[R_3] - \Delta H_f[R_1R_2R_3]^+$$
$$+ (T + U)\{[R_1R_3]^+ + [R_1]^+ + [R_2] + [R_3]\} - U[R_1R_3^+]$$

Now $T\{[R_1R_2]^+ + [R_3]\} + \{[R_1]^+ + [R_2]\}$ is the sum of the kinetic energies released in the two stages of mechanism (A) $= T_A$: Similarly $T\{[R_1R_3]^+ + [R_2]\} + T\{[R_1] + [R_2]\}$ is the corresponding total production of kinetic energy for route (B) $= T_B$. Thus,

$$\text{A.P. } [R_1^+]_A - \text{A.P. } [R_1^+]_B = T_A - T_B + U_A\{[R_1]^+ + [R_2] + [R_3]\}$$
$$- U_B\{[R_1]^+ + [R_2] + [R_3]\}$$

Thus,
$$2.95 - 2.56 = 0.55 - 0.17 + U_A - U_B$$

whence $U_A = U_B$ within experimental error.

Since all the excess energies in U_A and U_B are positive and independent quantities and their difference, within experimental error, is zero, it is reasonable to suppose that the independent components must also be zero. It would be remarkable if U_A and U_B were exactly equal, in view of the substantial differences in kinetic energies released, unless the individual excess energies were all zero.

Whilst this does not constitute a rigid demonstration that internal energies may be ignored in evaluating energy balances based on widths of metastable peaks, it may be said to point strongly in that direction in the present case, which may be taken as fairly typical of a fragmentation involving rearrangement and low frequency factor.

This may be interpreted either on the basis of statistical considerations or on that of a potential energy surface.

10   TRANSLATIONAL-ROTATIONAL ENERGY TRANSFER AND TUNNELLING MECHANISMS

A statistical equipartitioning of energy released between rotational and translational degrees of freedom would imply fluctuations in small ions that ought to be apparent in the slope of the shoulders of metastable peaks. In some cases, e.g. pyrazine, the peaks are gaussian and do not seem to imply large fluctuations. Thus, any rotational energy must be well defined by the elementary dynamics of fragmentation.

$(p = (^r\!/_2) \sin \theta)$

Suppose the hydrogen atom in methanol is released with velocity v at some angle θ relative to the $H\!-\!C\!\equiv\!O^+$ fragment. Then to conserve angular momentum

$$m_H p v = m_0 \omega (r/2)^2 + m_{CH} \omega (r/2)^2$$

where ω = angular velocity and we take the centre of mass as situated midway between the heavy atoms.

Rotational energy of $H\!-\!C\!\equiv\!O^+$ is $\frac{1}{2}(m_0 + m_{CH})(\omega^2 r^2)/4$. Translational energy of $H = \frac{1}{2}m_H v^2$, whence

$$\frac{\text{Rotational}}{\text{Translational}} = \frac{(m_0 + m_{CH})\,\omega^2 r^2}{4m_H v^2}$$

But

$$(m_0 + m_{CH}) \frac{\omega r^2}{4} = m_H p v$$

$$(m_0 + m_{CH}) \omega^2 r^2 = \frac{16 (m_H p v)^2}{(m_0 + m_{CH}) r^2}$$

Thus,

$$\frac{\text{Rotational}}{\text{Translational}} = \frac{4 (m_H p v)^2}{m_H v^2 (m_0 + m_{CH}) r^2}$$

$$= \frac{4 p^2 m_H}{(m_0 + m_{CH}) r^2} = \frac{\sin^2 \theta}{29}$$

if m_0, m_{CH}, m_H, are given their appropriate values.

There appears not to be any direct mechanism for generating angular momentum as such except through the intermediary of releasing an atom or group of atoms that can exert a couple on the resulting ion species. Conservation of angular momentum about the mass centre is otherwise rigidly enforced exactly as for the corresponding linear momentum as measured by a broadened metastable peak. There is nothing corresponding to the mechanism of energy transfer between different internal degrees of freedom. Consequently, the statistical equipartitioning of energy cannot apply to rotational and translational energy. The symmetry elements associated with invariance of the Hamiltonian operator for translation and rotation lead to the sharpness of the associated quantum numbers. Nevertheless, it is of interest to consider how rotational energy would be equipartitioned if the operators could be perturbed in such a way as to permit this.

Suppose the total energy released E comprises ε in the form of rotational energy and $(E - \varepsilon)$ in translational energy. Let there be L rotations and $(N - L)$ translational modes.

Then,

$$\varepsilon_j = \frac{lj (lj + 1) h^2}{8 \Pi^2 I} \qquad \varepsilon_j = \frac{n_j^2 h^2}{8 m a^2}$$

$$j = 1 \cdots L \qquad\qquad j = L + 1 \cdots N$$

The total number of distributions of energy[48]

$$= \int\int \cdots \int \frac{d\varepsilon_1 d\varepsilon_2 \cdots d\varepsilon_L}{\dfrac{d\varepsilon_1}{dl_1} \dfrac{d\varepsilon_2}{dl_2} \cdots \dfrac{d\varepsilon_1}{dl_L}} \int\int \cdots \int \frac{d\varepsilon_{L+1} d\varepsilon_{L+2} \cdots d\varepsilon_N}{\dfrac{d\varepsilon_{L+1}}{dn_{L+1}} \dfrac{d\varepsilon_{L+2}}{dn_{L+2}} \cdots \dfrac{d\varepsilon_N}{dn_N}}$$

$$\sum_{j=1}^{L} \varepsilon_j \leqslant \varepsilon \qquad\qquad \sum_{h=L+1}^{N} \varepsilon_j \leqslant E - \varepsilon$$

and, in view of

$$\frac{d\varepsilon_j}{dl_j} = \sqrt{\frac{h^2\varepsilon_j}{2\Pi^2 I_j}},$$

$$\frac{d\varepsilon_j}{dn_j} = \sqrt{\frac{h^2\varepsilon_j}{2ma^2}} = \left(\frac{2\Pi^2 I_j}{h^2}\right)^{L/2} \left(\frac{2ma^2}{h^2}\right)^{(N-L)/2} \times$$

$$\times \int (\varepsilon_1\varepsilon_2 \cdots \varepsilon_L)^{-1/2} \, d\varepsilon_1 d\varepsilon_2 \cdots d\varepsilon_L \int (\varepsilon_{L+1}\varepsilon_{L+2} \cdots \varepsilon_N)^{-1/2} \, d\varepsilon_{L+1} \cdots d\varepsilon_N$$

Using Dirichlet's integral:

$$I = \iint \cdots \int x_1^{i_1-1} x_2^{i_2-1} \cdots x_n^{i_n-1} \, dx_1 dx_2 \cdots dx_n$$

$$\left(\sum_{i=1}^{n} x_i \leqslant F\right)$$

$$= \frac{\Gamma(i_1)\,\Gamma(i_2) \cdots \Gamma(i_n)}{\Gamma(i_1 + i_2 + \cdots i_n)} F^{i_1 + i_2 + \cdots + i_n}$$

we find for the number of configurations:

$$\left(\frac{2\Pi^2 Ij}{h^2}\right)^{L/2} \left(\frac{2ma^2}{h^2}\right)^{(N-L)/2} \frac{[\Gamma(\frac{1}{2})]^N}{\Gamma(L/2)\,\Gamma\left(\frac{N-L}{2}\right)} \, \varepsilon^{L/2} \cdot (E - \varepsilon)^{(N-L)/2}$$

which is maximized when:

$$d/d\varepsilon \, [\varepsilon^{L/2} \, (E - \varepsilon)^{(N-L)/2} = 0$$

or

$$\varepsilon/E = L/N$$

For the ion involved in the methanol metastable, $H-C=\overset{+}{O}{}^{\diagup H}$

$$N = 5, \quad L = 2.$$

and $\frac{2}{5}$ of the energy released would be rotational. Obviously, fluctuation effects would be considerable because the above energy factor is not critically dependent on choice of ε/E. The sharp steep-sided peaks observed in many metastables (e.g. pyrazine ion losing HCN) would not occur. And of course, we could not explain the methanol results on such a basis.

A careful study of certain metastable peaks (2) in methanol arising from the reactions:

$$H-\dot{C} = O^+-H \rightarrow H-C\equiv O^+ + H\cdot -0.19 \, eV$$

and

$$\begin{array}{c} H \\ \\ H \end{array}\!\!\!\!\!C = O^+ \!-\! H \rightarrow H \!-\! C \equiv O^+ + H_2(\text{or } 2H \cdot) - 1.42 \text{ eV}$$

and deuterated analogues has shown that, notwithstanding the very considerable differences in exothermicity, consistent results for zero-point energies of CH and OH bonds can be obtained. The general approach here has been one of looking for internal consistency between similar reactions, and has been equally rewarding as that of the parallel reaction method which, as already remarked, helped to consolidate views on conservation of vibrational energy. These exothermic metastable ion decompositions are of special interest since a calculation of relative bond strengths in the ion from the widths of the metastable peaks accompanying the decompositions gives the apparent result that the bond strength of $O^+\!-\!H$ is greater than that of $O^+\!-\!D$ whilst that of $(C\!-\!D)$ is stronger than $C\!-\!H$. Whilst the latter is understandable the former is not, since we expect the greater zero-point energy for an atom of smaller mass will result in a lowering of bond strength. (see Figure 4 for a schematic energy diagram).

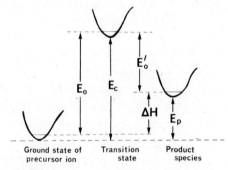

Figure 4 Schematic representation of the energy relationships for secondary ion dissociations of methanol
E_0 = forward activation energy from zero point energy level; E_c = classical activation energy; E_0' = reverse activation energy; ΔH = overall endothermicity

Whilst some simple classical explanation such as one invoking rotational energy might be imagined, this proves untenable for three main reasons: (i) the consistency achieved for two reactions of very different exothermicity, which would be most unlikely if a rotational component were involved, (ii) the $\dfrac{\text{rotational}}{\text{translational}}$ energy factor is $\approx \dfrac{4p^2m}{r^2\,(m_0 + m_{CH})}$ and is twice as big for deuterium as for hydrogen; this would imply an apparent weakening of any bond on deuteration because less energy is available for release as directly measurable kinetic energy, (iii) the absolute magnitude of the rotational energy for a reaction which is only 0.19 eV exothermic must be negligible compared with the observed difference in bond strengths.

Thus it appears that this type of approach can provide valuable information on rotational energy transfer, and this complements the similar information on vibrational energy from the method of parallel reactions.

Whilst the observation remains something of an anomaly we are still not sure whether this is a real anomaly insusceptible of explanation, or whether it might be due to the shape of the potential barrier and the capacity of a hydrogen atom to tunnel through it. Whilst tunnelling is considered to be a real phenomenon in solution chemistry it normally involves hydrogen transfer between two distinct species in close proximity. In principle we see no reasonable objection to an intramolecular and unimolecular type of tunnelling, when the barrier is due to an internal rearrangement of the single species.

Certainly, the additional excitation in $\overset{+}{OD}$ can be related to shape factors for the potential energy barrier and modified rate constants of the type:

$$k^1 = v \left(\frac{E - E_0}{E} \right)^{s-1} \exp \left\{ \frac{-2\Pi (E_0 - E_0^1)}{h v_t^1} \right\}$$

The tunnelling factor is based on a parabolic barrier as discussed by Bell[49].

A feature of the rotational energy component is that it is generally small (i) if the relative mass of the particle eliminated with reference to the mass of the residual system is small, (ii) if the potential surface has an axis of symmetry. Thus when $\theta = 0$ there is no rotational energy component (for $\sin^2 \theta/29 \rightarrow 0$). Likewise in the elimination of C_2H_2 from benzene, with the assumed form of the activated state of the ion the preserved vertical plane of symmetry implies that angular momentum is not generated.

Quite clearly, erroneous results can arise if the rotational energy component is ignored and thermochemical expressions are used which assume that all energy generated is the kinetic energy component observed as a broadening of the metastable peak. For example, replacement of hydrogen by deuterium should increase the rotational contribution by a factor of two.

In the methanol investigation we chose to ignore the rotational energy because our results were confirmed by consistency for $31^+ \rightarrow 29^+$, $30^+ \rightarrow 29^+$, involving very different amounts of kinetic energy. If the rotational energy component had been important it would have been of very different magnitude in the two cases. This might suggest that both hydrogens in $H\!-\!C\!=\!\overset{+}{O}H$ are directed along the symmetry axis or that, at least, this is so in the activated state if not in the lower energy state of the ion.

11 INTERNAL ROTATION

A clear case of a metastable transition free from rigid body rotation of separating fragments is that of the ketone dimer tetramethyl-1-3-cyclo-butanedione which is characterized by a metastable at $m^* = 50.5$ due to the simultaneous loss of two carbon monoxide molecules[50].

The transition state will no doubt retain two planes of symmetry at right angles and, assuming that an eclipsed form of $C_2(CH_3)_4$ is produced, the initial point group D_{2h} is preserved throughout the fragmentation process. There is no off centre force tending towards rotation of the tetramethyl-ethylene residue. On the other hand internal rotations of the pairs of methyls can occur to produce a staggered configuration. They may proceed by con- and dis-rotatory motions, and herein occur some interesting differences between photochemical and mass spectral mechanisms[51]. But as Turro *et al.* point out, the similarities are more conspicuous than the differences. Nevertheless, the breadth of the metastable and the amount of internal energy will be dependent upon the stereochemistry of the fragmentation. Again there are two principal co-ordinates x and y and, in principle, an appropriate mechanism to give energy interchange into internal vibrational and rotational modes. We may wish to ascertain whether the internal rotation can involve an appreciable internal energy component. Again, it appears that the kinetic shift (the excess energy above the minimum needed to result in dissociation) is sufficiently small in the case of a metastable ion that it does not include sufficient energy to set up vibrations in the products. Or in other words, the rotation in the transition state is insufficient in energy to produce an excited vibrational quantum in the final form of the product ion.

12   ANOMALOUS APPEARANCE POTENTIAL OF METASTABLE PEAKS IN AROMATIC NITRO COMPOUNDS

Reference has already been made to the anomalous appearance potential for the metastable elimination of neutral nitric oxide from the parent ion of nitrobenzene[14].

Appearance potential measurements for the metastable peak corresponding to loss of NO (or the anion NO^-) were made using the metastable mode of the M.S. 9 instrument and log abundance method of Lossing.[52] The normal daughter ion $(p - 30)^+$ had appearance potential lower than that of the corresponding metastable by some 2.4 eV. It seemed likely that a large part of the discrepancy might be due to the electron affinity of NO, assuming that an electron pair production were involved, viz:

$$p + e \rightarrow (p - 30)^+ + NO^- + e$$

Gaines and Page[53], however, used a magnetron method and found a value for electron affinity of 1.1 eV. Theoretical assessment might be based on the S.C.F. treatment of the NO molecule due to Brion and Moser[54].

If we designate the ground state configuration of the neutral $NO^{\cdot}$ molecule by:

$$KK\,(z\sigma)^2\,(y\sigma)^2\,(x\sigma)^2\,(w\pi)^4\,(v\pi)$$

that of the anion can be written:

$$KK\,(z\sigma)^2\,(y\sigma)^2\,(x\sigma)^2\,(w\pi)^4\,(v\overset{+}{\pi})\,(v\overset{-}{\pi}) \qquad {}^1\Sigma^+,\ {}^3\Sigma^-$$

$$KK\,(z\sigma)^2\,(y\sigma)^2\,(x\sigma)^2\,(w\pi)^4\,(v\pi)^2 \qquad {}^1\Delta$$

By Hund's rule, ${}^3\Sigma^-$ should be the ground state corresponding to the paramagnetic ground state of O_2 which is isoelectronic.

Using the tabulated values of coulomb, exchange and hybrid integrals[54], and the expressions for electron affinity, e.g. $A = -\varepsilon_i - J_{ij} + K_{ij}$, the calculated electron affinities turn out as follows:

${}^3\Sigma^-$	${}^1\Delta$	${}^1\Sigma^+$
-5.85 eV	-7.31 eV	-8.77 eV

These values are large and negative because the atomic integrals are overestimated if Slater orbitals are used[55].

A possible interpretation of the appearance potential measurements could conceivably take the view that the activated neutral state involved is the triplet ground state of nitrophenol and that this is the species which dissociates into a pair of ions, one of which is the NO^- anion in the $^3\Sigma^-$ state. But it is now believed that a chemical reaction rather than a triplet state of nitrophenol is primarily involved.

As we are neither too certain of the proper ground state of NO^- nor of the proper interpretation of the S.C.F. evaluation in absolute terms, little real evidence for pair production could be provided. The possibility was convincingly disproved by two independent criteria: (i) peaks tended to increase for several minutes after sample pressures had become constant and (ii) accurate mass measurement. A clear indication of a chemical reaction producing aniline was provided by the latter. This is discussed in detail in a recent paper[14]. The various considerations leading to this condition provide a good example of the combined use of careful measurement and observation combined with theoretical assessment. Whilst quantum chemistry is unable to provide more than a partial idea of the properties of metastable ions, it appears that the combined use of such powerful techniques as high resolution can often lead to positive conclusions.

13 VARIATION OF THE ENERGY RELEASED IN A META-STABLE DECOMPOSITION AS A FUNCTION OF LIFETIME

If a metastable ion decomposed always with release of the same amount of kinetic energy, then the sides of the metastable peak would be as steep as those of a normal mass peak. The fact that most metastable peaks are more diffuse than this suggests that there is, in fact, release of a range of kinetic energies in all such decompositions. But as we have seen, statistical distributions of rigid-body energy components would produce extremely diffuse peaks.

In the case of the rearrangement reaction in which pyrazine dissociates to eliminate HCN,

$$C_4H_4N_2^+ \rightarrow C_3H_3N^+ + HCN$$

the metastable peak is approximately gaussian and the reverse activation energy is close to zero[3]. Thus there should be no problems concerning components of either vibrational or rotational energy. We can attribute the width to the fact that dissociation at a given rate requires a definite but small excess energy usually called the "kinetic shift". It is apparently too small to leave the products in an excited state but varying amounts can enter the dissociation co-ordinate at the instant of crossing the energy barrier.

Using the simplest formulation of the unimolecular rate dependence on energy:

$$k(E) = v \left(\frac{E - E_0}{E} \right)^{s-1}$$

where the kinetic shift $= E - E_0$ and $s =$ effective number of vibrational oscillators. Thus:

$$\frac{\text{kinetic shift}}{\text{total internal energy}} = \frac{E - E_0}{E} = \left(\frac{k}{v} \right)^{(1/s-1)}$$

or

$$\frac{\text{kinetic shift}}{\text{activation energy}} = \frac{E - E_0}{E_0} = \left(\frac{(k/v)^{1/(s-1)}}{1 - (k/v)^{1/(s-1)}} \right)$$

The average unimolecular rate of decomposition

$$k(E) = \frac{v(s-1)}{E^{s-1}} \int_0^{(E-E_0)} (E - E_0 - y)^{s-1} \, dy$$

where y is that part of the kinetic shift that contributes to a direct release of kinetic energy.

Using the relation for effective mass

$$m^* = \frac{m_2^2}{m_1} \left(1 + \frac{\mu T}{eV} + 2 \sqrt{\frac{\mu T}{eV}} \right)$$

where $\mu = (m_1 - m_2)/m_2$ it then follows directly that:

$$T = E_1 + E_0 (1 - 2^{1/(s-2)}) \frac{(k/v)^{1/(s-1)}}{1 - (k/v)^{1/(s-1)}}$$

where T is the release of kinetic energy corresponding to half-height on the gaussian peak; $E_1 =$ reverse and $E_0 =$ forward activation energy.

Using an experimental technique in which the normal breadth of a peak can be compared with the metastable gaussian an extrapolation to zero width normal peak can be made.

In common with the earlier studies of nitronaphthalene and methanol this method can provide further valuable information on properties of the activated ions. In particular, the observation that the reverse activation energy is close to zero enables one to determine that 7 oscillators are effective in the pyrazine rearrangement and that the frequency factor $v = 1.4 \times 10^{12} \sec^{-1}$. This latter is low compared with a normal vibrational frequency as might be expected for a rearrangement involving entropy of activation.

It is gratifying that the width of a gaussian metastable peak can be quite well interpreted on the basis that rotational energy is not involved and that vibrational energy is conserved between reactant ion and product species.

Table 3 shows forward and reverse activation energies (eV) evaluated according to the above-mentioned procedure.

Table 3   Forward and reverse activation energies for the decomposition $C_4H_4N_2^{+\cdot} \rightarrow C_3H_3N^{+\cdot} + HCN$ in pyrazine

v (sec^{-1})	$s = 24$		$s = 8$		$s = 5$	
	E_0 (eV)	E_1	E_0	E_1	E_0	E_1
10^{14}	2.41	-0.0291	4.29	-0.0035	13.78	$+0.0059$
10^{12}	1.51	-0.0241	2.02	-0.0024	4.26	$+0.0061$
10^{10}	0.74	-0.0133	0.86	-0.0004	1.26	$+0.0065$
10^{8}	0.27	-0.0019	0.28	$+0.0039$	0.31	$+0.0082$

References

1. J. H. Beynon, R. A. Saunders, and A. E. Williams (1965) *Z. Naturforsch.*, **20a**, 180.
2. J. H. Beynon, A. E. Fontaine, and G. R. Lester (1968) *J. Mass Spec. and Ion Physics*, **1**, 1.
3. J. H. Beynon, J. A. Hopkinson, and G. R. Lester (1968) *J. Mass Spec. and Ion Physics*, **1**, 343.
4. G. R. Lester (1965) *Mass Spectrometry*, ed. R. I. Reed, Acad. Press, London and New York.
5. M. Born (1956) *Physics in My Generation*, Pergamon Press, London and New York (p. 127).
6. J. S. Griffith (1961) *The Theory of Transition—Metal Ions*, Cambridge.
7. H. Eyring, J. Walter, and G. E. Kimball (1944) *Quantum Chemistry*, J. Wiley and Sons, Inc., New York and London.
8. C. A. Coulson and K. Zalewski (1962) *Proc. Roy. Soc.* **A268**, 437.
9. G. G. Hall and R. D. Levine (1966) *J. Chem. Phys.* **44**, 1567.
10. R. D. Levine (1966) *J. Chem. Phys.* **44**, 2029, 2035, 2046, 3597.
11. J. Hamilton (1959) *The Theory of Elementary Particles*, Oxford Univ. Press, p. 212.
12. O. K. Rice (1961) *J. Phys. Chem.* **65**, 1588.
13. R. D. Levine (1964) *Bull. Soc. Chim. Belges*, **73**, 447.
14. J. H. Beynon, J. A. Hopkinson, and G. R. Lester (1969) *J. Mass Spec. and Ion Physics*, **2**, 291.

15. H.Budzikiewicz (1969) *Zeit. Analytische Chemie*, **244**, 1.
16. C.E.Melton, M.M.Bretscher, and R.Baldock (1957) *J. Chem. Phys.*, **26**, 1302.
17. J.H.Beynon, A.E.Fontaine, and G.R.Lester (1968) *Chem. Comm.*, 265.
18. A.S.Newton, A.F.Sciamanna, and R.Clampitt (1967) *J. Chem. Phys.*, **46**, 1, 1779.
19. L.B.Loeb (1927) *Kinetic Theory of Gases*, McGraw-Hill, New York, p. 462.
20. J.H.Beynon, B.E.Job and A.E.Williams (1966) *Zeit. Naturforsch.*, **21a**, 210.
21. R.J.Flurry, Jr. (1968) *Molecular Orbital Theories of Bonding in Organic Molecules*, Edward Arnold Ltd., London.
22. C.A.Coulson and R.Daudel (1956) *Dictionary of Values of Molecular Constants*.
23. T.E.Peacock (1959) *J. Chem Soc.*, 3645.
24. F.W.McLafferty and T.Wachs (1967) *J. Am. Chem. Soc.*, **89**, 5043.
25. I.Howe and D.H.Williams (1968) *J. Am. Chem. Soc.*, **90**, 5461.
26. A.Mandelbaum and K.Biemann (1968) *J. Am. Chem. Soc.*, **90**, 2975.
27. H.A.Jahn and E.Teller (1937) *Proc. Royal Soc.*, **A1**, 161, 220.
28. R.A.Saunders and A.E.Williams, *Advances in Mass Spectrometry*, (1966) Vol. **3**, W.L.Mead, Editor, Inst. of Petroleum, London, p. 681.
29. R.S.Ward, R.G.Cooks, and D.H.William (1969) *J. Am. Chem. Soc.*, **91**, 2727.
30. N.H.March (1957) *Advances in Physics*, **6**, 1.
31. N.H.March (1961) *Proc. Roy. Soc.* **A261**, 119.
32. A.H.Wilson (1953) *The Theory of Metals*, Cambridge U. Press, p. 163.
33. H.S.Carslaw and J.C.Jaeger (1947) *Conduction of Heat in Solids*, Oxford.
34. G.Mie (1908) *Ann. Phys.* **25**, 377.
35. G.Spiteller and M.Spiteller–Friedman (1965) *Liebigs Ann. Chem.* **690**, 1.
36. V.Santoro and G.Spadaccini (1969) *J. Phys. Chem.* **73**, 462.
37. K.Hirota, Y.Niwa, and M.Yamamoto (1969) *J. Phys. Chem.* **73**, 464.
38. G.R.Lester (1968) *Modern Aspects of Mass Spectrometry*, ed. R.I.Reed, Plenum Press.
39. G.G.Hall (1964) *Bull. Soc. Chim. Belge*, 305.
40. R.C.Dougherty (1968) *J. Am. Chem. Soc.*, **90**, 5780.
41. G.R.Lester (1963) *Brit. J. App. Phys.* **14**, 414.
42. Ch.Ottinger (1965) *Z. Naturforsch.* **20a**, 1229.
43. R.McWeeny and T.E.Peacock (1957) *Proc. Phys. Soc.* **A70**, 41.
44. S.Glasstone, K.J.Laidler, and H.Eyring (1941) *The Theory of Rate Processes*, McGraw-Hill, New York and London, p. 97.
45. P.E.Charters and J.G.Polani (1962) *Disc. Far. Soc.* **33**, 107.
46. F.T.Wall, L.A.Hiller jr. and J.Mazur (1958) *J. Chem. Phys.*, **29**, 255.
47. J.Perrin (1919) *Ann. Phys. IX*, **11**, 1.
48. H.M.Rosenstock (1952) *Thesis*, University of Utah.
49. R.P.Bell (1959) *Trans. Far. Soc.* **55**, 1.; R.P.Bell (1959) *The Proton in Chemistry*, Methuen, London.
50. N.J.Turro, D.C.Neckers, P.A.Leermakers, D.Seldner and P.D'Angelo (1965) *J. Am. Chem. Soc.*, **87**, 4097.
51. R.B.Woodward and R.Hoffmann (1965) *J. Am. Chem. Soc.*, **87**, 395.
52. F.P.Lossing, A.W.Tickner and W.A.Bryce (1951) *J. Chem. Phys.*, **19**, 1254.
53. A.F.Gaines and F.M.Page (1966) *Trans. Far. Soc.* **62**, 3086.
54. H.Lefebvre-Brion and C.M.Moser (1966) *J. Chem. Phys.*, **44**, 2951.
55. R.Pariser (1953) *J. Chem. Phys.*, **21**, 568.

The mass spectrometry of ferrocenes and related complexes*

GREGOR A. JUNK and HARRY J. SVEC

Institute for Atomic Research and Department of Chemistry
Iowa State University, Ames, Iowa 50010 USA

INTRODUCTION

The number of publications (see bibliography) dealing with some phase of the mass spectrometry of metallocenes† and related organometallic complexes is indicative of an ever increasing interest. The foundation for this interest is the first reported preparations of ferrocene [FeCp$_2$ where Cp designates the cyclopentadienyl radical, (C$_5$H$_5$)] by Kealy and Pauson[1] in 1951 and Miller *et al.*[2] in 1952. Several reports concerned with the "sandwich" structure[5-7] illustrated in Figure 1, the bonding[3,8,10] and the chemistry[17a] of ferrocene and other MCp$_2$ metallocenes followed very soon. These early results have since been refined, expanded and documented by others as discussed in recent textbooks[33b,76c].

Although the volatility of ruthenocene prompted its use for the investiga-

---

* Work was performed in the Ames Laboratory of the U.S. Atomic Energy Commission. Contribution No. 2695.

† Throughout the text the term metallocene is used to refer to all complexes where the bonding to the metal includes use of the π bonds of a cyclic ligand. Thus MgCp$_2$, MnCp$_2$, PrCp$_3$ etc. which are more accurately described as cyclopentadienides because of their considerable ionic bonding are included here as metallocenes. Even those complexes with π alkyl, π arene and other similar ligands are loosely referred to as metallocenes or metallocene-like complexes.

tion of the isotopic abundance of Ru metal in 1953,[9] the real impetus for studying metallocenes by the mass spectrometric method, was supplied by the classic work of Friedman *et al.* in 1955[11]. This report resulted from a systematic investigation of the 70 eV mass spectra, the ionization potentials and the ionic bond energies of the then almost complete catalogue of available complexes. The report however did not precipitate any immediate

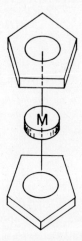

Figure 1　Pentagonal antiprism "sandwich" structure for MCp_2 complexes

feverish activity partially due to the incomplete development of the direct insertion sampling technique which is needed for many substituted metallocenes. In 1956, McLafferty[13] predicted a bright future in this area of mass spectrometry which seemed doomed when only one report[19] appeared in the next six years. Reed and Tabrizi[20] in 1963 reported the general features of mass spectra of ten mono-substituted ferrocenes and this report marked the beginning of the fulfillment of McLafferty's early prediction.

A rather narrow beginning in their mass spectrometry by Cais *et al.*[28,29] quickly developed into a broad investigation of a variety of substituted complexes. Other investigators, too numerous to cite individually are listed in the bibliography and/or Table 1 of Appendix A. In Table 1, the "Measurements" column is intended to present a quick indication of the type and completeness of the studies of the listed complexes. The classifications and the order of listing the complexes, while admittedly cumbersome, should be useful for locating individual complexes which have been studied. The table

Table 1 Summary of the literature on the mass spectrometry of ferrocenes and related complexes

Complex[a]	Measurements[b]	Refs.[c]
MCp		
M = Th	MS, MW, S	89
MCp$_2$		
M = Fe	AP, BE, IM, IP, m*, MS, MW, R, S	*11*, 13, 19, 23, 30, 42, 46, 47, 48, 70, 93
M = Ru	A, AP, BE, IP, MS, MW, S	9, *11*, 46
M = Os	IP, MS, MW, S	46
M = V; Cr; Mn	AP, BE, IP, MS, MW, S	11, 46
M = Co	AP, BE, IP, MS, MW, S	*11*, *46*, 48
M = Mg	AP, BE, IP, MS, MW, S	11
M = Yb	AP, BE, IP, MS, MW, S	84
M = Ni	AP, BE, IM, IP, MS, MW, S	*11*, 23, *46*, 47, 68
MCp$_3$		
M = Ti; Pr; Ho; Lu; Yb; Sm; Nd	AP, BE, IP, MS, MW, S	83, 84
M = As; Sb	MS, S	84
M = Tm	IP, MS, MW, S	84
M = U	BE, MS, S	80
MCp$_4$		
M = Th; U	AP, BE, IP, MS, MWt, S	80, 83
M(Ar)$_2^*$		
BzMCp M = Mn; Cr	AP, BE, IP, m*, MS, MW, R	42, 48, *63*, 78
M = Cr; Ar = 1,3,5,(CH$_3$)$_3$C$_6$H$_3$, (CH$_3$)$_6$C$_6$, C$_6$H$_5$C$_6$H$_5$	AP, BE, IP, m*, MS, MW, R	86

Table 1 *(cont.)*

Complex[a]	Measurements[b]	Refs.[c]
BzCrC$_6$H$_5$C$_6$H$_5$	AP, BE, IP, m*, MS MW, R	86
CpMCh M = V; Cr; Mo	BE, m*, MS, MW, S	*63*, 76, 78
ChCrC$_7$H$_{10}$	MS, MW	75
CpMC$_8$H$_{12}$ M = In; Rh	m*, MS, MW, R, S	78
CpNiC$_{13}$H$_{17}$	m*, MS, MW, R, S	78
MBz$_2$ M = V; Cr	AP, BE, IP, M*, MS, MW, R	48, *63*, 86

Cp$_2$MH$_x$		
Cp$_2$TcH	BE, IP, MS, MW, R, S	46, 53, 54
Cp$_2$ReH	BE, IP, MS, MW, R, S	11, 46, 53, 54
Cp$_2$WH$_2$	BE, IP, MS, MW, R, S	46

Halocarbon		
CpFe(CO)$_2$X X = CH$_2$C$_6$F$_5$; C≡CCF$_3$; cis − CF$_3$CH=CH	MS, MW, S, VC	55a, 76c
CpM(CO)$_2$CF=CFCF$_3$ M = Fe; Ru	MS, MW, S, VC	91
CpTi(C$_6$F$_5$)X X = Cl; C$_6$H$_5$	MS, MW, S, VC	76d
CpFeC$_5$H$_4$R R = C$_6$F$_5$; COC$_6$F$_5$; CH$_2$C$_6$H$_5$; C$_6$F$_4$OCH$_3$(p)	MS, MW, S, VC	90
Fe(C$_5$H$_4$R)$_2$ R = same as above	MS, MW, S, VC	90
C$_4$F$_5$Fe(CO)$_2$Cp	MS, MW, S, VC	76b
C$_5$F$_6$ClFe(CO)$_2$Cp	MS, MW, S, VC	76b
C$_6$F$_9$M(CO)$_2$Cp M = Fe; Ru	MS, MW, S, VC	76b

Table 1 *(cont.)*

Complex[a]	Measurements[b]	Refs.[c]
Halogen ligand		
$CpFe(CO)_4Br$	AP, BE, MS	41
$CpMo(CO)_3Br$	AP, BE, MS	41
Pd and Rh allylic complexes	MS	74
Cp_2MX_2 M = Ti; Zr; Hf X = Cl; Br; I	MS, MW, R, VC	81, 32
Cp_2MCl_2 M = Ti; Zr	AP, BE, IP, m*, MS MW	79
$[Cp_2M(Cl)]_2O$ M = Ti, Zr	MS, VC	31, 32
$CpTi(OEt)_{3-n}Cl_n$ n = 0 to 3	AP, BE, IP, MS, MW	26
$CpM(CO)_2X$ M = Ru; Fe X = Cl; Br; I	BE, MS, MW, R, S	76
Cp_3UCl	BE, MS, MW, R, S	80
Cp_3MF M = Th; U	IP, MS, MW, S	83
Cp_2YbCl_2	IP, MS, MW, S	83
CP_3MI M = Th; U	MS	83
$ArM(CO)_x$*		
$CpV(CO)_4$	AP, BE, IP, m*, MS, MW, S	**33**, 88
$CPMn(CO)_3$	AP, BE, IP, m*, MS, MW, S	33, 64
$CpCo(CO)_2$		33, 48
$CpMn(CO)_2L$ L = Cy(C_5H_8); Cy(C_7H_{12}); Cy(C_8H_{14}) nor (C_7H_8); nor C_7H_{10} maleic anhydride; isonitrile; amine; phosphine; sulfoxide	IP, m*, MS, MW R	64

Table 1 *(cont.)*

Complex[a]	Measurements[b]	Refs.[c]
CpMn(CO)L L = butadiene	IP, m*, MS, MW R	64
CpMn(CO)$_2$PX$_3$ X = i-C$_3$H$_7$; i-OC$_3$H$_7$ C$_6$H$_5$; H; Cl; Br OC$_6$H$_5$	IP, m*, MS, MW R	88
(CO)$_3$MnC$_5$H$_4$COR R = H; Me; OH; CHN$_2$; CH$_2$CH$_2$CO$_2$H; CH$_2$CH$_2$CH$_2$CO$_2$H; CH$_2$COCO$_2$Et	m*, MS, MW, R, S	73
(CO)$_3$MnC$_5$H$_4$R R = CH$_2$OH; CH(OH) (CH$_3$) CH$_2$CO$_2$Et; (CH$_2$)$_3$C(OH)Ph$_2$; (CH$_2$)$_4$CO$_2$H, CH(CH$_3$)CHCO$_2$MeCOOH; C = CHCOOH CH$_2$CH$_2$COOH	m*, MS, MW, R, S	28, 73

$$\underset{\displaystyle R=R'=H;\ R=\ R'=\ Me;\ H\quad H;\ Me\quad Et;\ Me}{(CO)_3Mn\overset{\displaystyle R'}{C}=CH-CO_2R}$$

	m*, MS, MW, R, S	73

R =	R' =	m*, MS, MW, R, S	28
CH$_3$	OH		
CH$_3$	OD		
CH$_3$	NH$_2$		
CH$_3$	ND$_2$		
COOH	OH		
COOCH$_3$	OCH$_3$		

| (CO)$_3$CrC$_6$H$_5$X
X = H; CH$_3$; C$_6$H$_5$; F;
I; OH; NH$_2$; COCH$_3$
OCOCH$_3$; CO$_2$R, CH$_2$CO$_2$Et;
CH=CHCO$_2$Et | M*, MS, MW
R | 48, 85 |

Table 1 *(cont.)*

Complex[a]	Measurements[b]	Refs.[c]
$(CO)_3CrC_6H_5X$ X = H; F; Cl; Br, I; $COCH_3$; CO_2CH_3; C_6H_5; CH_3; $Si(CH)_3$; OCH_3; $C(CH_3)_3$, NH_2	AP, BE, IP, m* MS, MW, R, S	48, *87*
$(CO)_3Cr[1,3,5-C_6H_3(CH_3)_3]$	AP, BE, IP, m* MS, MW, R, S	87
$CpFe(CO)_4Br$	AP, BE, IM, MS, MW	41
$CpMo(CO)_3Br$	AP, BE, IM, MS, MW	41

Mono-substituted ferrocenes

$CpFeC_5H_4XR$

XR = CN; $CH(OH)CH_3$; C_6H_5 $NHCO_2CH_3$; C_6H_4Cl; $C_6H_4NO_2(p)$; SO_2NH_2; $CO(CH_2)_3CO_2CH_3$; CH_3; C_2H_5; $CO(CH_2)_3COOH$; $CH_2C_6H_5$; C_2H_5; $C_6H_4N_2O$; CH_2OH; $CH(OH)CH_2CH = CH_2$; $-\overline{CH(CH_2)_2COO}$; Cis and trans $-\overline{CHCH_2CHCH_2OH}$; cis and trans $\overline{CHCH_2CHCOOEt}$	MS, MW, R, S	*20, 25*, 30, 40
XR = CN; CH_2OH; $COCH_3$; OCH_3; $C_6H_3(CH_3)_2$; $C_6H_4CH_3(p)$; $C_6H_4Cl(p)$ $C_6H_4Cl(m)$; $CONHCH_3$; $CH_2NC_5H_{10}$	AP, BE, IP, m*, MS MW, R, S	92, 93, 94
X = (CH=CH) and R = COOH; COMe; cis COOMe; trans COOMe; $CH_2CH_2O(CO)Me$ X = (C_6H_5COO) and R = H; CH_3; C_2H_5; n–C_3H_7 XR = $C\equiv CCO_2H$; $CH=CCl_2$ $CH=CBr_2$	m*, MS, MW, R, S	77
X = C_6H_4 and R = $O(CO)CH_3$; $O(CO)Et$; $COCH_3$	L, m*, MS, MW, R, S	57
XR = $(CH_2)_n$ OH $n = 1, 2, 3, 4$; $CH(OH)CH_3$; $CH(OH)C_6H_5$	L, m*, MS, MW, R, S	36
X = CO and R = OH; OD; CH_3; $NHCH_3$; C_6H_5; $C_6H_4OCH_3(p)$; OCH_3	L, m*, MS, MW, R, S	22

Table 1 *(cont.)*

Complex[a]	Measurements[b]	Refs.[c]
$XR = Si(C_6H_5)_3$; $Si(CH=CH_2)(CH_3)_2$ $Si(CH_3)_3$; $Si(OEt(CH_3)_2)$; $B(OH)_2$	LVMS, MW, QA	19, 34
$(CH_2)_nCOOCH_3$ (n = 0–5)	L, MS, R	58

Hetero di-substituted ferrocenes
$RC_5H_4FeC_5H_4R'$

$R = R' = (CH_2=CH)$	L, m*, MS, MW, R, S	69
$R = R' = CH_2N(CH_3)_2$; Cl C_2H_5; n-C_4H_9; COC_4H_9; $Si(CH_3)_3$; $Si(OEt)(CH_3)_2$; $Si(CH=CH_2)(CH_3)_2$; COC_3H_7; $Si(OMe)(CH_3)_2$	LVMS, QA, MS, MW	19, 34
$R = R' = CH_2OH$; $CH(OH)CH_3$; $R = CH_2NMe_2$; $R' = C(C_6H_5)OH^d$	L, m*, MS, MW, R, S; QA, MS; MW, S	36 34

Polysubstituted ferrocenes

	MS, MW	82
$Fe[C_5(CH_3)_5]_2$	MS, MW	49
	MS, MW	34

Cyclic di-substituted ferrocenes

	L, MS, R, S	52
	L, MS, R, S	37
	L, MS, R, S	35, 37

Table 1 *(cont.)*

Complex[a]	Measurements[b]	Refs.[c]

Ferrocenophanes *f*

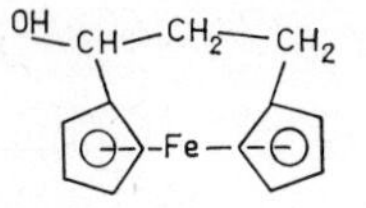

L, MS, MW, R, S   36

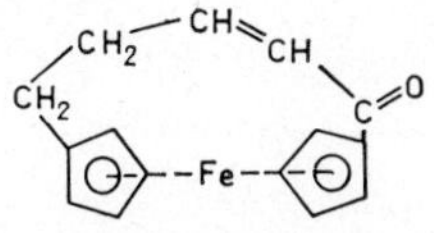

MS, MW, S   66

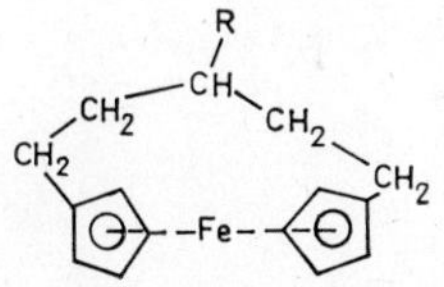

MS, MW, S   66

Polymetal systems

Complex[a]	Measurements[b]	Refs.[c]
$C_6H_3[1,2,4(Fc)_3]^e$	MW	51
$C_6H_3[1,3,5(Fc)_3]$	MW	51
$(Cp_2TiCN)_3$	MS, MW	55
$(Cp_2TiNCS)_3$	MS, MW	55
$Cp_4Rh_3(H)$	MW	50
$Cp_2M_2(CO)_4$	BE, MS, MW	76a
M = Ru; Fe		
$CpRuRe(CO)_7$	MS	76a
$CpRuCo(CO)_6$	MS	76a
$(CF_3C \equiv CH)(NiCp)_2$	MS; MW; S	76c
$[CpM(CO)_2]_2$	AP, BE, MS, MW, S	41
M = Fe; Ni		
$[CpMo(CO)_3]_2$	AP, BE, MS, MW, S	41
$(Cp_2YbNH_2)_2$	IP, m*, MS, MW, S	83
$(C_{10}H_8FeHg)_x$	SSMS	21

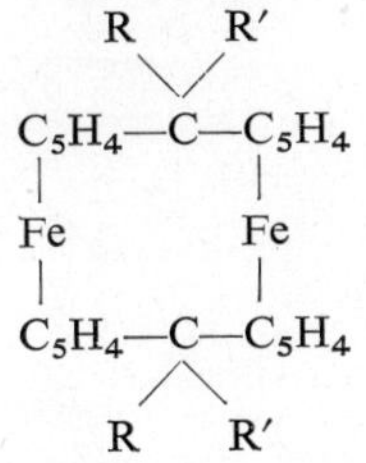

m*, MS, MW, S   43, 45, 65

Table 1 *(cont.)*

Complex[a]	Measurements[b]	Refs.[c]
$R = H$; $R' = CH_3$		
$R = R' = CO$		
$R = R = H$; CH_3		
$(C_{17}H_{13}Fe)_2$	MS, MW, S	29
$(Fc)_2$; $(Fc)_3$	MW	19, 30
7 complexes with an unsaturate, carbonyls and dimetal combinations of Fe, Mn; Fe, Cr; Mn, Mn; Cr, Cr	MS	28

Miscellaneous		
Cp_2TiNCO	Im, MS, MW	55
$CpRh(C_2H_4)_2$	BE, m*, MS, MW, R, S	78
$CpPtCH_3$	BE, m*, MS, MW, R, S	78
$CpPdC_3H_5$	BE, m*, MS, MW, R, S	78
$CpPdC_{10}H_{12}OCH_3$	BE, m*, MS, MW, R, S	78, 70
Tl	MS, MW, S	52
$Cp_2Zr(H)BH_4$	MS, MW	56

[a] The centre heading (e.g. MCp_2) classifies the type of complex and the designation in this column is used in conjunction with the center heading whenever possible to identify the complex.

[b] Abbreviations used for the type of measurements made on the listed complex are:

A – used for isotope abundance measurements.

AP – fragment ion appearance potentials are reported.

BE – ionic and some neutral bond dissociation energies reported or available from reported data.

HRMS – high resolution mass spectra used to establish empirical formula IM – ion molecule reactions observed.

IP – ionization potentials reported.

L – isotope labelling used to help establish fragmentation mechanisms.

LVMS – low voltage mass spectrometry.

m* – observed metastables used to help establish fragmentation mechanisms.

MS – either complete or partial mass spectra reported.

MW – molecular formula established by observation of parent ion.

QA – quantitative analysis prospects discussed.

R – fragments formed by rearrangements are discussed.

S – mass spectra used to differentiate structural isomers or as supplementary empirical evidence for structure of complex.

SSMS – sprak source mass spectra reported.

VC – valency change concept used to partially interpret the observed mass spectra.

[c] Where multiple references are listed, the references in bold are of primary importance either for the number of complexes studied or for the discussion of the mass spectra.

[d] This complex is homo-disubstituted and is the only one reported in the literature.

[e] Fc is an abbreviation for ferrocenyl radical and is $(C_5H_5FeC_5H_4)$.

[f] The complexes reported in Ref. 43, 45 and 46 listed in the polymetal class are also ferrocenophanes.

* (Ar) is a general dsignation for an unsaturate ligand such as benzene, cyclopentadienyl, cycloheptatrienyl etc.

is complete for MCp_x metallocenes and substituted ferrocenes, $CpFeC_5H_4XR^*$. Although not complete for metallocene-like complexes, a representative list is included. The reader is referred to the comprehensive review by Bruce[60] for other organometallic complexes. A partial review by Lewis[60a] and two short reviews by King[71,72] are also suggested as supplementary surveys.

The similarities which exist in the 70 eV mass spectra of some classes (as classified in Table 1) of metallocenes is both striking and useful. The following mass spectral discussions will concentrate on those general fragmentation features which seem to permeate the entire spectra of each class of complexes without giving attention to isolated, albeit interesting features which have not yet been tested on enough similar complexes. In this respect, the discussion is analogous to those which developed in the early 1950's regarding general features of hydrocarbon mass spectra and in the early 1960's regarding natural product mass spectra. Some mass spectral features which are particularly interesting or unique und show promise of future development have been emphasized. Thus, the subject of rearrangements, where a systematic and comprehensive attack is obviously needed but not yet completed is considered separately.

Also considered separately are ionization potentials and bond dissociation energies. It is revealing to note that recent books and reviews on metallocenes make extensive mention of U.V., I.R., X-ray, NMR, kinetic and thermochemical data. Mass spectral citations and discussions are noticeably absent except for infrequent reference to the work of Friedman *et al.* (1955). One explanation for this omission is probably the complexity and incompleteness of mass spectral information which precludes the development of simple empirical rules of fragmentation which can be directly related to the metallocene structure. A second reason may be that scientists outside the field of mass spectrometry are not interested in mass spectral data *per se*, except for molecular weight determinations where only the molecule-ion is of interest. Fragmentation novelties not firmly related to structural features merely result in excess tabulations of *m/e* and intensity data. It seems probable that outside interest is currently centered around the energy of the most loosely bound electron in the metallocene system and the metal-ligand bond dissociation energy. This belief dictated the inclusion of separate sections on ionization potentials and bond energies.

---

* $CpFeC_5H_4$· radical is often called ferrocenyl and abbreviated Fc so that substituted ferrocenes may be designated by FcXR or FcR.

MASS SPECTRA—GENERAL FEATURES

Cyclopentadienyl metal complexes [MCp$_x$]

Mass spectral data is available for the entire first row transition metal bis-cyclopentadienyl complexes as well as $RuCp_2$, $OsCp_2$ and $MgCp_2$. The ferrocene ($FeCp_2$) mass spectrum shown in Figure 2 is representative of this class of metallocenes.

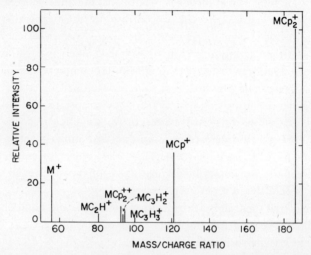

Figure 2   Representative mass spectra (70 eV) of MCp$_2$ complexes. The line diagram here is for ferrocene

The fragmentation pattern which is relatively simple has the following gross characteristics:

1. an intense parent ion peak, usually the base peak;

2. a large peak due to the loss of one Cp radical;

3. very little fragmentation of the Cp rings;

4. a detectable metal ion (M^+) peak;

5. a detectable doubly charged parent ion;

6. low intensity rearrangement peaks resulting from loss of the neutral metal atom;

7. more than 90% of the observed ions contain the metal atom;

8. almost negligible ion currents due to ligands only.

The mass spectra of the bis-cyclopentadienides such as $MnCp_2$, $MgCp_2$ and $YbCp_2$ have the same general features but with subtle differences such as a lower parent ion intensity and more fragmentation of the Cp rings.

The MCp_3 and MCp_4 cyclopentadienides have the same general features as the MCp_2 cyclopentadienyls and cyclopentadienides but with a much reduced or negligible M^+ intensity.

Other cyclic unsaturated ligands [M(Ar)₂]

Various ligand combinations of benzene, cyclopentadienyl, cycloheptatrienyl etc. have been coupled with the transition elements to form $M(Ar)_2$ complexes where Ar designates some aromatic type ligand. The mass spectral characteristics of these complexes are the same as the MCp_2 complexes with the following *exceptions*:

1. lower parent ion intensity;
2. more ions resulting from fragmentation of the ligand;
3. larger ion currents due to charged ligands;
4. larger M^+ (metal) ion currents.

Features 3 and 4 may well be due to the lower thermal stabilities of $M(Ar)_2$ complexes with consequent thermal decomposition in the ion source and hence are artifacts.

Mono-substituted metallocenes [CpFeC₅H₄XR]

Numerous 70 eV spectra of ferrocenyl complexes are available for establishing a general fragmentation scheme. Analogous complexes with metals other than iron, for which spectra are not currently available, should follow the same fragmentation scheme. Any significant differences would be due to the manner in which the metal participates in the decomposition mechanisms. The mass spectra of $CpFeC_5H_4COCH_3$ is used in Figure 3 to represent this class of metallocenes. As in the case of the MCp_2 complexes, the spectra of the substituted derivatives are simple with the following features:

1. a parent ion peak which is usually the base peak;
2. a detectable, sometimes intense peak due to loss of a Cp ring;
3. peaks due to fragmentation of some part or all of the XR substituent;
4. a detectable metal ion (M^+) peak;

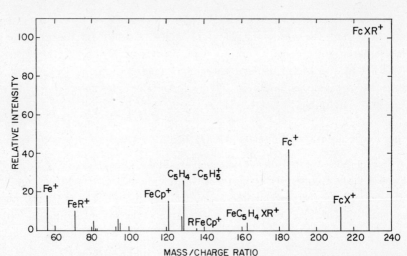

Figure 3   Representative mass spectra (70 eV) of mono-substituted metal complexes (CpMC$_5$H$_4$XR). The line diagram here is for CpFeC$_5$H$_4$COCH$_3$

5. an intense MCp$^+$ peak;

6. detectable doubly charged parent ion peaks;

7. much less than 90% of the observed ions contain the metal atom;

8. extensive rearrangement peaks dependent on the size and elemental composition of XR.

Whether these general features apply to substituted cyclopentadienides cannot be established owing to a lack of mass spectral data. It is assumed that they apply although minor adjustments may be necessary.

A better understanding of the fragmentation behavior of substituted ferrocenes is gained from consideration of Figure 4 and Table 2[28,94] where the letter labelled ion intensities are tabulated. From even these limited data, it seems as if attempts to generalize the spectra of these complexes are a bit naive. Note that X and R can drastically affect the observed ion intensities. Nevertheless it is believed that this general pattern will prevail for most substituted metallocene spectra. If the R group is highly electronegative (such as a halogen) then the valency change concept[33,34] may dictate the fragmentation scheme and large ion currents due to the formation of MR$^+$ or the elimination of neutral MR from the parent ion may be observed. If R is a bulky unsaturate such as phenyl then the charge resulting from ionization

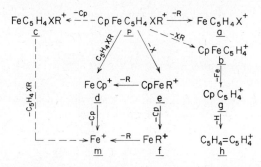

Figure 4 Partial fragmentation scheme for $CpFeC_5H_4XR^+$ where $X = O$, $R = CH_3$; $X = CH_2$, $R = OH$ and $CH_2NC_5H_{10}$; $X = CO$, $R = CH_3$; OH; $NHCH_3$; C_6H_5; $C_6H_4OCH_3$ and OCH_3. Metastables were observed only for the solid arrow reactions

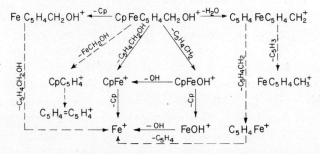

Figure 5 Partial fragmentation scheme for $CpFeC_5H_4CH_2OH^+$ (see also Table 2)

Table 2 Mass spectra (70 eV) of monosubstituted ferrocenes, $CpFeC_5H_4XR$ (see Fig. 4)

Complex		Fragment ion intensities									
X =	R =	P	a	b	c	d	e	f	g	h	m	Refs.
CO	CH_3	100	12	52	5	17	0.2	10	26	6	33	22, 92
CO	$NHCH_3$	100	8	7	34	23	12	2	10	4	21	22, 92
CO	C_6H_5	100	2	4	3	6	3	9	4	2	19	22
CO	$C_6H_4OCH_3$	100	1	2	3	8	1	3	3	3	13	22
CO	OCH_3	100	4	4	1	26	44	1	6	3	36	22
CO	·OH	87	0	0	15	0	100	23	2	2	89	22
CH_2	OH	75	2	1	9	13	100	21	1	1	14	92
CH_2	NC_5H_{10}	75	89	0	1	50	2	1	1	1	17	92
O	CH_3	100	23	1	1	41	1	1	2	1	14	92

may be largely delocalized with a consequent lowering of the number of metal containing fragment ions and a complimentary increase in ligand (and ligand fragment) ion currents.

A partial fragmentation scheme of $CpFeC_5H_4CH_2OH$ is presented in Figure 5 to clarify the fragmentation rules given above, particularly rules 3 and 8. A strong tendency to fragment the XR group exists and the $CpFeOH^+$ rearrangement ion current has supplanted the parent ion (see Table 2) as the base peak.

Poly-substituted ferrocenes

Although some work[19,34,36,49,69,86] has been published dealing with the mass spectra of homo- and hetero di-substituted and a few other poly-substituted complexes the spectral information is sparse and no general rules can be applied. Sufficient partial spectra are reported however which suggest that the fragmentation of this class of metallocenes is complex and not subject to simple empirical rules. If rules which allow one to draw structural conclusions do exist then these will be established only after the systematic accumulation and analyses of many more spectra than are now available. In this regard, the preparation of a variety of alkyl substituted ferrocenes by Bublitz[82] is intriguing and the mass spectral measurements from such a class of complexes may be even more structurally useful than the infra-red measurements[82]. This is not meant to suggest that no structural information is currently available for this class of complexes. Indeed the presence of a $m/e = 121$ ion current distinguishes homo from heterosubstituted ferrocenes[28] similar to the presence of an absorption in the 9–10 μ region in infrared spectra[14]. Nevertheless, a largely untapped potential for correlating mass spectral data with structure exists for this class of metallocenes.

A special kind of di-substituted ferrocene, in which the substituent is a link between the two Cp rings, are the ferrocenophanes studied by Watts[43,44,65]. The fragmentation scheme for these complexes as reproduced by Bruce[60] is informative. This scheme suggests that a wealth of structural information will be available once a comprehensive list of spectra are catalogued.

Aromatic-carbonyl-metal combinations [(Ar) N(CO)$_x$]

The mass spectra of CpM(CO) complexes[33] where M = V, Mn and Co and $x = 4$, 3 and 2 respectively are representative of this class of metallocene-like complexes. The general features are:

1. almost universal observation of large parent ion;
2. an intense $M(Ar)_n^+$ ion current;
3. negligible $M(CO)_x^+$ ion currents;
4. detectable M^+ ion current;
5. no fragmentation of Ar ligand prior to loss of all CO groups;
6. fragmentation of the Ar ligand of the $M(Ar)^+$ ion.

Halogen-aromatic-metal combinations

The fragmentation scheme for bis[chloridodi(cyclopentadienyl)zirconium] oxide[31,32] given in Figure 6 may eventually provide the clue for characterizing this general class of complexes. The ability of the metal atom to change valence appears to dictate many of the observed peaks in the spectrum. However, this valence change concept has not yet been applied to a sufficient number of spectra so that a useful and consistent pattern is established. Only three positive features seem to characterize these spectra:

1. the parent ion is usually observed;
2. MX_n^+ ions are observed where X = halogen;
3. MX_n neutrals are often eliminated from the parent or one of the fragment ions.

Halogen ligands

When halogens are substituted for hydrogens on one or both of the aromatic ligands the mass spectra are similar to the unsubstituted complexes. Thus the general features are the same as those given for MCp_x and $M(Ar)_2$ complexes, although some differences, attributable to a weakening of the metal-ligand bonds are observed. Bruce[55a,76a-d,90,91] has recently published several spectra which demonstrate this weakening effect and the general tendency of these complexes to form MX^+ or MX_2^+ (X = halogen) ion currents and/or ions resulting from elimination of neutral MX or MX_2 from the parent or one of the fragment ions.

Present utility

The almost universal observation of a parent ion, usually an intense peak, make the mass spectral measurements extremely useful for confirmation of

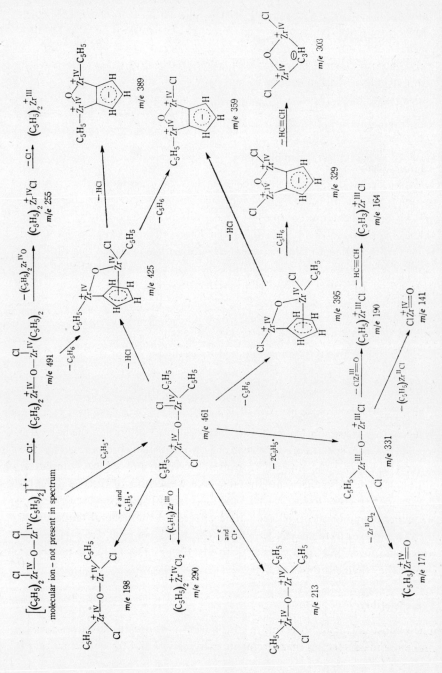

Figure 6 Fragmentation scheme for bis [chloridodi(cyclopentadienyl)zirconium] oxide ions, $Cp_2ClZr\text{-}O\text{-}ZrClCp_2^+$ (see also Ref. 32)

the molecular weight and the empirical formula if isotopic analysis or high resolution mass spectrometry are employed. The intensity of the parent ion peak furthermore gives at least a clue to the structure and bonding. Rearrangements, which are discussed below, have already been used for distinguishing several structural isomers and orientations[35,36,37,40,57,58]. Empirical structural relationships based on mass spectral data are beginning to be established. An estimate of the relative strength of metal-ligand bonds is sometimes available from comparisons of the spectra of similar complexes.*

Future promise

It is expected that the number of mass spectra published will continue to grow primarily because of the molecular weight information provided. The present tendency to utilize mass spectra solely for this purpose and the consequent reporting of only partial spectra may seriously limit the development of empirical structural relationships but this will be partly offset by the availability of spectral tape catalogues and an expected systematic attack by a few investigators on structurally similar and authentic samples. These attacks will undoubtedly include rearrangement reactions, labelling experiments and metastable ion current studies† with more emphasis placed on the lower intensity ion currents. Attempts to correlate conclusions based on mass spectral results with those drawn from I.R. and U.V. spectroscopy, charge-transfer spectra, kinetics, ion stabilities, X-ray structure determinations, and thermochemical measurements will hopefully be made. Certainly the ionization potentials of more complexes will be measured. The use of mono-energetic electron beams for these measurements will help the accuracy of the measurements and remove some of the discrepancies. There appears to be no real justification for the present lack of photon impact measurements since these complexes are easily sampled and yield usually intense parent ions. Appearance potentials of more fragment ions will be used to measure bond dissociation energies and the thermal decomposition of many complexes, presently considered to be detrimental for mass spectral studies, may be an asset for the determination of the ionization potentials of many radicals.

---

* This approach has serious inherent difficulties and should be applied only if appearance potential (A.P.) data is not available. Conclusions based on this approach should be viewed with guarded skepticism.

† It should be noted that a valid general feature of the mass spectra of all metallocenes is the observation of large numbers of metastable ion currents of appreciable intensity.

This will require deconvolution of the ionization efficiency curves but the initial success of Müller[63] with this procedure is encouraging. Finally, some one may make a serious attempt to accumulate and compare the mass spectra and ionization potentials of substituted cyclopentadienes[27] and other ligands with the mass spectra of complexes containing these ligands.

REARRANGEMENTS

The R designation in Table 1 (Appendix A) catalogues those publications and complexes for which some aspects of rearrangement reactions are discussed. Individual consideration of all these reports is beyond the scope of this chapter and not enough studies have been reported to allow very many useful generalizations. It is obvious that rearrangements are more extensive and more important with metallocenes than one is led to believe from a study of the literature to the present time. While several investigators have already attacked this problem[22,28,35,36,37,57,58,64a,69,73,76,77] the solution is far from complete. Some results do suggest that the observed rearrangements have analogies in organic mass spectrometry. For example the formation of the $CpFeOH^+$ rearrangement ion from CpFeXOH (where X = CO, CH_2 etc.) molecules may be considered to occur through a concerted mechanism which utilizes the pseudo-four membered cyclic transition state shown in Figure 7. Pseudo-five, -six and -seven or larger membered rings may be

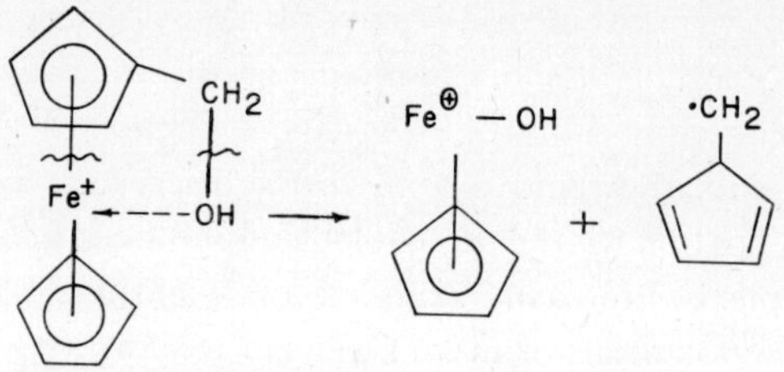

Figure 7 Pseudo four-membered ring cyclic transition state for the formation of the $CpFeOH^+$ ion from $CpFeC_5H_4OH^+$

employed as X changes from CH_2 to $(CH_2)_2 \cdots (CH_2)_n$. The ability to form a rearrangement ion may well be dependent on the metal, whose effect has not yet been studied, and on the stability of the R anion[73].

Another possible explanation for the observation of many rearrangement ions may be a relationship between the stability of the positive ion formed in the mass spectrometer and the stabilized α-metallocenylcarbonium ion concept[18,38] used to explain chemical reactions of metallocenes. A model description of the bonding which explains the metal atom participation in stabilizing α-metallocenylcarbonium ions is shown in Figure 8. The expansion of the metal-ligand bonding to include the α position may be extended to the β position of substituted metallocenes[38]. It seems plausible that a ring shift[18] and possibly even a ring tilt[18a] occurs during or after ionization in the mass spectrometer so that the metal atom is partially bonded to the α and β positions (for example, the CH_2 and OH groups in Figure 7) in the parent ions. Concerted bond ruptures and formation of a strong metal-OH bond would then explain the facile elimination of the (C_5H_3) neutral from the parent ion.

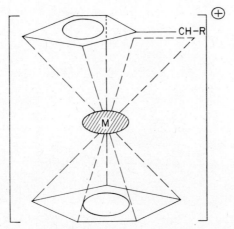

Figure 8   Stabilized α-metallocenylcarbonium ion

The molecular orbital calculations of Ballhausen and Dahl[18a] are also useful in formulating a reasonable mechanism for rearrangements where the metal atom is eliminated from metallocenes and substituted metallocenes to form ions such as $(C_5H_4 = C_5H_5)^+$. The theoretically predicted ability of the cyclopentadienyl rings in charged metallocenes to tilt up to 65 degrees out of a parallel plane with no significant decrease of the metal-ligand bond strength suggests that the activation energy for tilting is near zero. Thus one expects severe tilting of the rings for positively charged ions so that the rings are in a much more favourable position for forming a bond in a unimolecular rearrangement decomposition.

Table 3 Ionization potentials (I.P.) of metallocenes and related compounds*

Complex	I.P.^{a} Metal	I.P. Complex		ΔIP^b	Refs.
CpVCp	6.74	7.56	7.33	−0.6	11, 46
CpCrCp	6.76	6.91	6.26	+0.5	11, 46
CpMnCp	7.43	7.25	7.32	+0.1	11, 46
CpFeCp	7.90	7.05	7.15	+0.8	11, 46
		7.1	6.99		92, 47
		6.94^c	7.00^c		61, 61
CpCoCp	7.86	6.2	6.21	+1.7	11, 46
	5.95				48
CpNiCp	7.63	7.06	7.16	+0.4	11, 46
		6.75	6.8		47, 67
CpRuCp	7.36	7.82		−0.5	46
CpOsCp	8.5	7.59		+0.9	46
CpMgCp	7.64	7.76		−0.2	11
CpFeC$_5$H$_4$(CH)$_3$	7.90	6.76^c		+1.1	61
C$_5$H$_4$(CH$_3$)FeC$_5$H$_4$(CH$_3$)	7.90	6.65^c		+1.2	61
(C$_3$H$_7$)C$_5$H$_4$FeC$_5$H$_4$(C$_3$H$_7$)	7.90	6.85^c		+1.0	61
CpFeC$_5$H$_4$OCH$_3$	7.90	6.8		+1.1	92
CpFeC$_5$H$_4$CH$_2$NC$_5$H$_{10}$	7.90	6.9		+1.0	92
CpFeC$_5$H$_4$C$_6$H$_4$CH$_3$(p)	7.90	7.1		+0.8	92
CpFeC$_5$H$_4$CH$_3$(CH$_3$)$_2$	7.90	7.2		+0.7	92
CpFeC$_5$H$_4$C$_6$H$_4$Cl(m)	7.90	7.3		+0.7	92
CpFeC$_5$H$_4$C$_6$H$_4$Cl(p)	7.90	7.3		+0.6	92
CpFeC$_5$H$_4$CONHCH$_3$	7.90	7.2		+0.8	92
CpFeC$_5$H$_4$CH$_2$OH	7.90	7.3		+0.7	92
CpFeC$_5$H$_4$COCH$_3$	7.90	7.6		+0.3	92
CpFeC$_5$H$_4$CN	7.90	7.8		+0.1	92
CpFeCp	7.90	7.1		+0.8	92
CpMnBz	7.43	6.92	7.1	+0.5	63, 42
CpCrBz	6.76	6.13		+0.6	63
CpCrCh	6.76	5.96		+0.8	63
BzVBz	6.74	6.26		+0.5	63
CpVCh	6.74	7.24		−0.5	63
BzCrBz	6.76	5.91	7.0	+0.9	86, 42
		5.70			48
Cr[(CH$_3$)$_3$C$_6$H$_3$]$_2$	6.76	5.47		+1.3	86
Cr[(CH$_3$)$_6$C$_6$]$_2$	6.76	5.19		+1.6	86
BzCrC$_6$H$_5$−C$_6$H$_5$	6.76	5.94		+0.8	86
Cr(C$_6$H$_5$−C$_6$H$_5$)$_2$	6.76	5.87		+0.8	86

Table 3 *(cont.)*

Complex	I.P.[a] Metal	I.P. Complex		ΔIP[b]	Refs.
$TiCp_3$	6.82	6.47		+0.3	83
$NdCp_3$	5.7	8.30		−2.6	84
$SmCp_3$	5.6	8.01		−2.4	84
$PrCp_3$	5.46	7.68		−2.2	83
$HoCp_3$	5.89	7.46		−1.6	83
$TmCp_3$	5.81	7.43		−1.6	83
$YbCp_3$	6.2	7.30	7.72	−1.1	83, 84
$YbCp_2$	6.2	7.62		−1.1	84
$LuCp_3$	5.32	7.36		−2.0	83
$ThCp_4$	6.95	7.41		−0.5	83
UCp_4	6.50	6.08		+0.4	83
Cp_2TiCl_2	6.82	8.98		−2.2	79
Cp_2ZrCl_2	6.92	9.37		−2.5	79
Cp_3UF	6.08	7.53		−1.5	83
Cp_3ThF	6.95	8.06		−1.1	83
Cp_2YbCl_2	6.2	8.65		−2.5	83
$Cp_4Yb_2(NH_2)_2$	6.2	7.87		−1.7	83
$CpTc(H)Cp$	7.28	7.13		+0.1	46
$CpRe(H)Cp$	7.87	6.76		+1.1	46
$CpW(H_2)Cp$	7.98	6.49		+1.5	46
$ChCr(CO)_3$	6.76	7.10		−0.3	87
$CpCr(CO)_3$	6.76	7.30		−0.5	87, 85
$CpV(CO)_4$	6.74	8.2		−1.4	33
$CpMn(CO)_3$	7.43	8.3	8.12	−0.1	33, 88
$CpCo(CO)_2$	7.86	8.3	7.78	+0.1	33, 48
$(CO)_3CrC_6H_5F$	6.76	7.47		−0.7	87
$(CO)_3CrC_6H_5COCH_3$	6.76	7.44		−0.7	87
$(CO)_3CrC_6H_5COOCH_3$	6.76	7.41		−0.7	87
$(CO)_3CrC_6H_5Br$	7.76	7.40		−0.6	87
$(CO)_3CrC_6H_5Cl$	7.76	7.37		−0.6	87
$(CO)_3CrC_6H_5I$	6.76	7.36		−0.6	87
$(CO)_3CrBz$	6.76	7.30	7.39	−0.5	85, 48
$(CO)_3CrC_6H_5C_6H_5$	6.76	7.27	7.35[c]	−0.5	87
$(CO)_3CrC_6H_5CH_3$	6.76	7.19	7.39[c]	−0.4	87
$(CO)_3CrC_6H_5Si(CH_3)_3$	6.76	7.15			87
$(CO)_3CrC_6H_5OCH_3$	6.76	7.11	7.38[c]	−0.4	87
$(CO)_3CrC_6H_5C(CH_3)_3$	6.76	7.08			
$(CO)_3CrC_6H_5NH_2$	6.76	7.05			

Table 3 *(cont.)*

Complex	I.P.[a] Metal	I.P. Complex		ΔIP^b	Refs.
$(CO)_3CrC_6H_3(CH_3)_3$	6.76	7.05	7.33^c	-0.3	87
$(CO)_3CrC_6H_5N(CH_3)_2$	6.76	6.92	7.28^c		87
$(CO)_3CrC_6(CH_3)_6$	6.76	6.88	7.24^c	-0.1	87
$(CO)_3CrC_6H_4[N(CH_3)_2]_2$	6.76	6.46	7.14^c		87
$CpMn(CO)_2P(iC_3H_7)_3$	7.43	6.90	6.55	$+0.5$	88, 64
$CpMn(CO)_2P(C_6H_5)_3$	7.43	6.93		$+0.5$	88
$CpMn(CO)_2P(OC_3H_7)_3$	7.43	7.17		$+0.3$	88
$CpMn(CO)_2PH_3$	7.43	7.28		$+0.2$	88
$CpMn(CO)_2(POC_6H_5)_3$	7.43	7.40		$+0.0$	88
$CpMn(CO)_2PBR_3$	7.43	8.01		-0.6	88
$CpMn(CO)_2PCl_3$	7.43	8.12		-0.7	88
$CpMn(CO)_2NH(CH_3)_2$	7.43	6.55		$+0.9$	64
$CPMn(CO)_2CNC_6H_{11}$	7.43	7.01		$+0.4$	64
$CpH_5Mn(CO)_2$ Maleic anhydride	7.43	8.04		-0.6	64
$CpMn(CO)_2SO(CH_3)_2$	7.43	7.12		$+0.3$	64
$CpMn(CO)_2nC_7H_{10}$	7.43	7.19		$+0.2$	64
$CpMn(CO)_2nC_7H_8$	7.43	7.27		$+0.2$	64
$CpMn(CO)_2C_8H_{14}$	7.43	7.00		$+0.4$	64
$CpMn(CO)_2C_7H_{12}$	7.43	7.12		$+0.3$	64
$CpMn(CO)_2C_5H_8$	7.43	7.29		$+0.1$	64
(VCp)	6.74	7.8		-1.1	63
$(CrCp)$	6.76	6.7		$+0.1$	63
$(NiCp)$	7.63	7.8		-0.2	67
(VBz)	6.74	6.8		-0.1	63
$(CrBz)$	6.76	6.4		$+0.4$	63
$(MgCp)$	7.64	7.62		$+0.0$	11^d
$(MnCp)$	7.42	7.94		-0.5	11^d
$(FeCp)$	7.90	9.78		-1.9	11^d
$(CoCp)$	7.86	10.05		-2.2	11^d
(Cp)		8.69	8.56		16, 67
HCp		9.00			16
Bz		9.24			17
HCh		8.55			16
Ch		6.60			16
$Cp-Cp$		7.75			67

* Abbreviations—Cp = Cyclopentadienyl radical $(C_5H_5\,\cdot)$ or Cyclopentadienide ion $(C_5H_5)^-$; HCp = Cyclopentadiene (C_5H_6); Ch = Cycloheptatrienyl radical $(C_7H_7\,\cdot)$; HCh = Cycloheptatriene (C_7H_8); Bz = Benzene (C_6H_6).

[a] Values taken from Ref.17. All metal atom values are from spectroscopic measurements.

IONIZATION POTENTIALS

General features

A complete list of the electron impact measured ionization potentials (I.P.) of metallocenes and related complexes is tabulated in Table 3. The references from which I.P.'s were taken are recorded in the order of appearance of the values. Several I.P.'s calculated from charge-transfer spectra[61] are included for comparison purposes although in a few cases these are the only measurements available. The ionization potentials of several ligands and some (MCp) and (MAr) radicals are also listed at the end of the table. The spectroscopic values for the metal atoms and the I.P.'s of the complexes are compared in the (ΔIP) column for convenient reference.

It is immediately apparent that the ionization potentials of the metals and

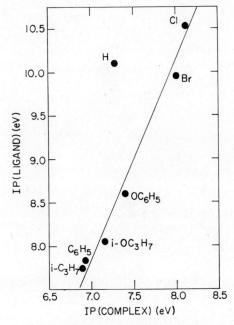

Figure 9 Correlation between the ionization potentials of the PX_3 ligands and the $CpMn(CO)_2PX_3$ complexes

[b] The ionization potential of the metal atom minus the ionization potential of the complex.

[c] Calculated I.P. from charge transfer (CT) spectra of complexes in polar solvents.

[d] No direct measurements available. These are the estimated values from Ref. 11 where only the MgCp value should be considered accurate. The values for MnCp, FeCp and CoCp are obviously high. See text for explanation.

metallocenes are nearly equal. This observation has lead to the frequent conclusion that ionization of organometallic complexes is from an orbital whose character is primarily metallic. The large number of positive ΔIP values for the covalently bonded complexes is significant. A casual explanation for this observation may be that the excess of electrons around the metal nucleus, when the bonds to the ligand are formed, elevates one of the "d" electrons into an orbital of higher energy than exists in the ground state metal atom. It is this electron which is removed on ionization instead of a bonding electron or one of the available electrons of the ligand, the energetics of which are expected to be much higher than the observed ionization potentials of the complexes.

In most cases the rather high negative ΔIP values for those metallocenes which are known to have appreciable ionic bonding is characteristic. The exceptions are UCp_4, $ThCp_4$, $TiCp_3$, $MnCp_2$ and $MgCp_2$.

Correlations

The ionization potentials of several metallocenes and metallocene-like complexes have been related to other measurements. Thus a relationship

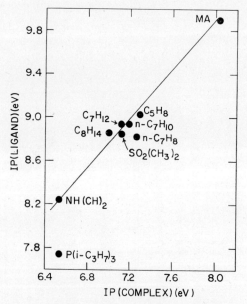

Figure 10   Correlation between the ionization potentials of the ligands (L) and the $CpMn(CO)_2L$ complexes

Table 4 Ionization potentials of substituted metallocenes related to the I.P. of the substituent, the C—O stretching frequency and the ν(CO) force constants[a]

Complex	I.P. of complex eV	I.P. of sub. eV	ν(CO) cm^{-1}		K mdyn/Å
CpMn(CO)$_3$	8.12	14.01	2028	1944	15.71
CpMn(CO)$_2$PCl$_3$	8.12	10.52	1994	1944	15.66
CpMn(CO)$_2$PBr$_3$	8.01	9.85	1994	1945	15.67
CpMn(CO)$_2$(POC$_6$H$_5$)$_3$	7.40	8.60	1968	1906	15.16
CpMn(CO)$_2$PH$_3$	7.28	10.10	1958	1899	15.02
CpMn(CO)$_2$P(iOC$_3$H$_7$)$_3$	7.17	8.05	1947	1881	14.80
CpMn(CO)$_2$P(C$_6$H$_5$)$_3$	6.93	7.83	1944	1883	14.79
CpMn(CO)$_2$P(iC$_3$H$_7$)$_3$	6.90	7.75	1932	1870	14.60
CpMn(CO)$_2$ Maleic anhydride	8.04	9.9	2012	1955	
CpMn(CO)$_2$C$_5$H$_8$	7.29	9.01	1950	1891	
CpMn(CO)$_2$C$_7$H$_{12}$	7.12	8.94	1950	1890	
CpMn(CO)$_2$C$_8$H$_{14}$	7.00	8.26	1951	1889	
CpMn(CO)$_2$nC$_7$H$_8$	7.27	8.73	1957	1898	
CpMn(CO)$_2$nC$_7$H$_{10}$	7.19	8.98	1952	1894	
CpMn(CO)$_2$SO(CH$_3$)$_2$	7.12	8.85	1949	1884	
CpMn(CO)$_2$CNC$_6$H$_{11}$	7.01	—	1944	1890	
CpMn(CO)$_2$NH(CH$_3$)$_2$	6.55	8.24[b]	1916	1840	
(CO)$_3$CrC$_6$H$_5$F	7.47	9.20			15.012
(CO)$_3$CrC$_6$H$_5$COCH$_3$	7.44	9.27			15.064
(CO)$_3$CrC$_6$H$_5$COOCH$_3$	7.41	10.0			15.100
(CO)$_3$CrC$_6$H$_5$Br	7.40	8.98			15.042
(CO)$_3$CrC$_6$H$_5$Cl	7.37	9.07			15.038
(CO)$_3$CrC$_6$H$_5$I	7.36	8.73			15.022
(CO)$_3$CrC$_6$H$_6$	7.30	9.25			14.852
(CO)$_3$CrC$_6$H$_5$C$_6$H$_5$	7.27	8.27			14.860
(CO)$_3$CrC$_6$H$_5$CH$_3$	7.19	8.82			14.774
(CO)$_3$CrC$_6$H$_5$SiCH$_3$	7.15	—			14.814
(CO)$_3$CrC$_6$H$_5$OCH$_3$	7.11	8.22			14.774
(CO)$_3$CrC$_6$H$_5$C(CH$_3$)$_3$	7.08	8.68			14.743
(CO)$_3$CrC$_6$H$_5$NH$_2$	7.05	7.70			14.615
(CO)$_3$CrC$_6$H$_3$(CH$_3$)$_3$	7.05	8.39			14.652

[a] CpMn(CO)$_3$PX$_3$ data from Ref. 88; CpMn(CO)$_2$X data from Ref. 64; (CO)$_3$CrX data from Ref. 87.
[b] Photoionization value – all others are electron impact measurements.

exists between the ionization potential of the ligand and the ionization potential of the complex[64,88] as shown in Figures 9 and 10 and tabulated as part of Table 4. This relationship is not universally observed however and has rather high error limits for some complexes (see PX_3 in Figure 9). The work of Foffani and Pignataro with substituted carbonyls[47,68a] suggests that this correlation is expected when the bonding of the ligand to the metal atom involves a high degree of σ character. The failure or success of the correlation may then be used as a vital clue to the type of bonding which exists in the complexes.

A second correlation exists between the ionization potentials of monosubstituted ferrocenes and the Hammett sigma (σ) values for the substituents. This relationship is plotted in Figure 11. Large uncertainties of the order of 0.1 eV for the I.P.'s and 0.1 unit for the σ values make this correlation somewhat uncertain. If however its validity is established from studies of additional compounds, the correlation may be used as supporting evidence for localized ionization on the metal atom since the substituent effect (σ) dictates the availability of electrons at the reaction site. The "site" in this case is the metal atom and the "reaction" is ionization. A recent report[85a] includes a

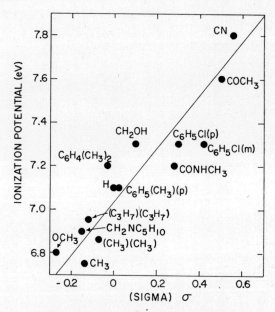

Figure 11   Correlation between Hammett sigma (σ) values for R and the ionization potentials of substituted ferrocenes (FcR)

similar correlation between the ionization potentials of several chromium hexacarbonyl derivatives and the Hammett σ values for the substituents.

A third correlation exists between the I.P.'s and (C—O) stretching frequency data tabulated in Table 4. A direct relationship between the ν(CO) force constants and the I.P.'s of the complexes[64,87,88] is evident in the plots in Figures 12 and 13. These results are to be expected if one resorts to the

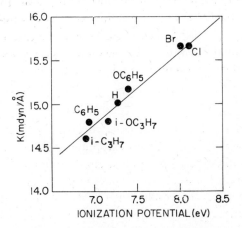

Figure 12 Correlation between the ionization potentials and the ν(CO) force constants for CpMn(CO)$_2$PX$_3$ complexes

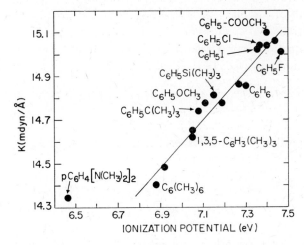

Figure 13 Correlation between the ionization potentials and the ν(CO) force constants for LCr(CO)$_3$ complexes

generally accepted practice of inversely relating $\nu(CO)$ measurements to the relative strength of the M—C bonds in metal carbonyls. The C—O and M—C stretching frequencies are considered complementary so that any observed decrease in $\nu(CO)$ is accompanied by an increase in $\nu(MC)$. This increase is directly related to an increase in the M—C bond strength. A π-bonding concept, in which one of the variables contributing to the strength of a metal-ligand bond is the availability of "d" orbital electrons from the metal for backbonding to the CO ligand, is then invoked. Any ligand effects which increase the availability of "d" electrons should increase the M—CO bond energy and simultaneously decrease the ionization potential. This model for ionization and bonding is consistent with the existing data.

Present utility

Ionization potentials are fundamental properties of any molecular system. Organometallic compounds are no exception and these measurements are useful for making estimates of bond strengths in addition to providing experimental information useful to the development of ligand field and molecular orbital theories. The tabulations (Tables 3 and 4) given here are evidence for a localized ionization model. The correlations which exist not only further the understanding of metallocene chemistry but are potentially useful for estimating the ionization potentials of complexes not yet studied or those which are difficult to measure. Three simple equations for calculating ionization potentials of complexes from the ionization potential of a single complex and other ancillary values can be derived from the data in Figures 9–13.

The first equation derived from Figures 9 and 10 utilizes ancillary values of the ionization potentials of the ligands and is

$$IP = IP_k + A \, (\text{Ligand } IP_{unk} - \text{Ligand } IP_k) \tag{1}$$

where subscripts "k" and "unk" refer to known and unknown and A is a constant the value of which is 0.44 and 0.88 for $CpMn(CO)_2PX_3$ and $CpMn(CO)_2L$ complexes respectively.

The second equation derived from Figure 11 utilizes ancillary Hammett σ values and is

$$IP = IP_k + A \, (\sigma_{unk} - \sigma_k) \tag{2}$$

where A is again a constant whose values is 1.22 for substituted[2] ferrocenes.

The third equation derived from Figures 12 and 13 utilizes ancillary $\nu(CO)$

force constants and is,

$$IP = IP_k + A\,(K_{unk} - K_k) \tag{3}$$

where A is again a constant, the value of which is 0.92 and 1.22 for $LCr(CO)_3$ and $CpMn(CO)_2PX_3$. respectively.

Future developments

The striking absence of reports of photon impact measurements should be remedied because of the demonstrated ease of sampling metallocenes and the gradual elimination of a presently exaggerated ion source contamination problem. These measurements will undoubtedly yield more accurate and probably more reproducible values. However the current need is for more ionization potential measurements rather than a resurgence of the controversy concerning the merits of photon impact, mono-energetic electron impact and conventional electron impact measurements. It is hoped that conventional electron impact measurements will remain acceptable since many laboratories are just beginning to acquire equipment for such measurements and past results suggest that valid and valuable conclusions may be drawn from these measurements. The more sophisticated measurements may be better reserved for periodic and representative testing of conclusions based upon more conveniently measured electron impact values.

The success of applying deconvolution techniques[63] to ionization efficiency data may provide the incentive for the measuring of the ionization potentials of many free radicals which are difficult to generate and maintain using a heated reactor removed from the ion source region. The expected large reproducibility limits for these radical I.P.'s, due to difficulties in the ionization efficiency curve extrapolation to onset values, is tolerable in view of the need for radical I.P.'s in calculating specific neutral dissociation energies which cannot be arrived at by any other method.

BOND DISSOCIATION ENERGIES

Ionic vs. neutral $\overline{D}$

The average dissociation energies determined from appearance potentials (A.P.) and I.P.'s are tabulated in Table 5 for the dissociation processes listed. The ionic values were calculated from ionization and appearance potential data taken from the listed references. Neutral values were calculated for all

Table 5   Comparison of ionic and neutral bond dissociation energies for metallocenes and related complexes in eV

Dissociation process[a]	$\bar{D}$ (neutral)[b]		$\bar{D}$ (ionic)[c]		ΔD[d]	Refs.
$MgCp_2 \to Mg + 2Cp$	$(3.4)^e$		$(3.3)^e$		+0.1	11,17
$MnCp_2 \to Mn + 2Cp$	3.1	$(3.3)^e$	3.1	$(3.4)^e$	+0.0	46,11,17
$VCp_2 \to V + 2Cp$	3.9	$(6.1)^e$	3.6	$(5.4)^e$	+0.3	46,11,17
$CrCp_2 \to Cr + 2Cp$	4.2	$(4.7)^e$	3.9	$(4.6)^e$	+0.3	46,11,17
$FeCp_2 \to Fe + 2Cp^f$	3.6	$(4.6)^e$	3.3	$(5.0)^e$	+0.3	46,11,17
$CoCp_2 \to Co + 2Cp$	4.2	$(4.2)^e$	3.4	$(5.0)^e$	+0.8	46,11,17
$NiCp_2 \to Ni + 2Cp^g$	3.2	$(3.4)^e$	3.0	$(3.6)^e$	+0.2	46,11,17
$RuCp_2 \to Ru + 2Cp$	4.1		4.4		−0.3	46,17
$VCp_2 \to VCp + Cp$	4.9	$(4.9)^e$	5.3	$(5.1)^e$	−0.4	46,11,63
$VCp \to V + Cp$			1.8	$(5.6)^e$		46,11
$CrCp_2 \to CrCp + Cp$	6.1	$(6.9)^e$	6.5	$(6.7)^e$	−0.4	46,11,63
$CrCp \to Cr + Cp$			1.8	$(2.6)^e$		46,11
$MnCp_2 \to MnCp + Cp$	3.2	$(3.3)^e$	3.8	$(4.0)^e$	−0.6	46,11,17
$MnCp \to Mn + Cp$			2.5	$(2.8)^e$		46,11
$FeCp_2 \to FeCp + Cp$	4.0	$(4.6)^e$	6.6	$(7.3)^e$	−2.6	46,11
$FeCp \to Fe + Cp$			0.6	$(2.7)^e$		46,11
$CoCp_2 \to CoCp + Cp$	4.0	$(4.1)^e$	7.8	$(8.0)^e$	−3.8	46,11
$CoCp \to Co + Cp$			0.7	$(1.9)^e$		46,11
$NiCp_2 \to NiCp + Cp$	4.8	$(4.9)^e$	5.4	$(5.6)^e$	−0.6	46,11,67
$NiCp \to Ni + Cp$			1.1	$(1.6)^e$		46,11
$MgCp_2 \to MgCp + Cp$	$(3.4)^e$		$(3.2)^e$		+0.2	11
$MgCp \to Mg + Cp$			3.4			11
$NdCp_3 \to Nd + 3Cp$	5.8		5.0		+0.8	84,17
$NdCp_3 \to NdCp_2 + Cp$			1.7			84
$NdCp_2 \to NdCp + Cp$			7.8			84
$NdCp \to Nd + Cp$			5.4			84
$SmCp_3 \to Sm + 3Cp$	5.5		4.7		+0.8	84,17
$SmCp_3 \to SmCp_2 + Cp$			2.0			84
$SmCp_2 \to SmCp + Cp$			4.4			84
$SmCp \to Sm + Cp$			7.8			84
$YbCp_3 \to Yb + 3Cp$	3.9	4.3	3.4	4.2	+0.5	84,83,17
$YbCp_3 \to YbCp_2 + Cp$	2.5	1.3	2.4	1.6	+0.1	84,83,17
$YbCp_2 \to YbCp + Cp$			2.1	3.1		84,83
$YbCp \to Yb + Cp$			5.6	7.0		84,83
$PrCp_3 \to PrCp_2 + Cp$			0.8			83
$PrCp_2 \to PrCp + Cp$			8.5			83
$HoCp_3 \to HoCp_2 + Cp$			1.5			83
$LuCp_3 \to LuCp_2 + Cp$			1.6			83
$LuCp_2 \to LuCp + Cp$			9.4			83

Table 5 *(cont.)*

Dissociation process[a]	$\bar{D}$ (neutral)[b]	$\bar{D}$ (ionic)[c]	$\overline{\Delta D}$[d]	Refs.
$ThCp_4 \rightarrow ThCp_3 + Cp$		0.3		83
$UCp_4 \rightarrow UCp_3 + Cp$		0.8		83
$UCp_3 \rightarrow UCp_2 + Cp$		7.8		83
$TiCp_3 \rightarrow TiCp_2 + Cp$		1.8		
$CrBz_2 \rightarrow Cr + 2Bz$[h]	2.0	2.4	-0.4	86,63,17
$CrBz_2 \rightarrow CrBz + Bz$	2.4	2.9	-0.5	63,86
$BzCr \rightarrow Cr + Bz$		2.0		63,86
$VBz_2 \rightarrow V + 2Bz$[i]	3.4	3.7	-0.3	63,17
$VBz_2 \rightarrow VBz + Bz$	4.2	4.2	0.0	63
$VBz \rightarrow V + Bz$		3.1		63
$BzCrCp \rightarrow Cr + Bz + Cp$	3.7	3.6	$+0.1$	63,17
$BzCrCp \rightarrow CrCp + Bz$	2.9	2.5	$+0.4$	63
$CrCp \rightarrow Cr + Cp$		4.6		63
$BzMnCp \rightarrow Mn + Bz + Cp$	3.3	3.6	-0.3	63,17
$BzMnCp \rightarrow MnCp + Bz$	1.5	2.5	-1.0	63,11
$MnCp \rightarrow Mn + Cp$		4.7		63
$ChCrCp \rightarrow Cr + Ch + Cp$	2.7	3.1	-0.4	63,17
$ChCrCp \rightarrow CrCp + Ch$	3.4	4.4	-1.0	63
$CrCp \rightarrow Cr + Cp$		1.8		63
$ChVCp \rightarrow V + Ch + Cp$	3.5	3.3	$+0.2$	63,17
$ChVCp \rightarrow VCp + Ch$	1.4	2.0	-0.6	63
$VCp \rightarrow V + Cp$		4.6		63
$BzCrC_6H_5C_6H_5 \rightarrow Cr + Bz + C_6H_5C_6H_5$	2.4	2.8	-0.4	86,17
$BzCrC_6H_5C_6H_5 \rightarrow CrC_6H_5C_6H_5 + Bz$		3.0		86
$CrC_6H_5C_6H_5 \rightarrow Cr + C_6H_5C_6H_5$		2.7		86
$Cr(C_6H_5C_6H_5)_2 \rightarrow Cr + 2(C_6H_5C_6H_5)$	2.8	3.2	-0.4	86,17
$Cr(C_6H_5C_6H_5)_2 \rightarrow CrC_6H_5C_6H_5 + C_6H_6$		3.7		86
$CrC_6H_5C_6H_5 \rightarrow Cr + C_6H_5C_6H_5$		2.8		86
$Cr[C_6H_3(CH_3)_3]_2 \rightarrow Cr + 2[C_6H_3(CH_3)_3]$	2.7	3.3	-0.6	86,17
$Cr[C_6H_3(CH_3)_3]_2 \rightarrow CrC_6H_3(CH_3)_3 + C_6H_3(CH_3)_3$		3.8		86
$CrC_6H_3(CH_3)_3 \rightarrow Cr + C_6H_3(CH_3)_3$		2.8		86
$Cr[C_6(CH_3)_6]_2 \rightarrow Cr + 2[C_6(CH_3)_6]$	3.6	4.4	-0.8	86,17
$BzCr(CO)_3 \rightarrow Cr + Bz + 3(CO)$	1.5	1.4	$+0.1$	85,17
$BzCr(CO)_3 \rightarrow CrBz + 3(CO)$	1.4	1.1	$+0.3$	85,63
$CrBz \rightarrow Cr + Bz$		2.6		85
$CpV(CO)_4 \rightarrow V + Cp + 4(CO)$	2.4	2.2	$+0.2$	33,17
$CpV(CO)_4 \rightarrow VCp + 4(CO)$	1.8	1.5	$+0.3$	33,63
$VCp \rightarrow V + Cp$		5.2		33

Table 5 *(cont.)*

Dissociation process[a]	$\bar{D}$ (neutral)[b]		$\bar{D}$ (ionic)[c]		$\Delta\bar{D}$[d]	Refs.
$CpMn(CO)_3 \rightarrow Mn + Cp + 3(CO)$	2.1		1.9		+0.2	33,17
$CpMn(CO)_3 \rightarrow MnCp + 3(CO)$	1.4		1.2		+0.2	33,11
$CpMn \rightarrow Mn + Cp$			3.9			33
$CpCo(CO)_2 \rightarrow Co + Cp + 2(CO)$	3.0		2.8		+0.2	33,17
$CpCo(CO)_2 \rightarrow CoCp + 2(CO)$	0.8	0.4	1.7	1.5	−0.9	33,62,11
$CpCo \rightarrow Co + Cp$			5.1			33
$V(CO)_6 \rightarrow V + 6(CO)$	1.5		1.4		+0.1	59
$Cr(CO)_6 \rightarrow Cr + 6(CO)^j$	1.4		1.4		0.0	59[k]
$Mo(CO)_6 \rightarrow Mo + 6(CO)^j$	1.8		1.7		+0.1	59[k]
$W(CO)_6 \rightarrow W + 6(CO)^j$	2.2		2.1		+0.1	59[h]
$Fe(CO)_5 \rightarrow Fe + 5(CO)^j$	1.5		1.4		+0.1	59[h]
$Ni(CO)_4 \rightarrow Ni + 4(CO)^j$	1.8		1.6		+0.2	59[h]

[a] All reactants and products are in the gaseous state.

[b] $\bar{D}$ (neutral) is used to designate average dissociation energy e.g. $\bar{D}$ for $MgCp_2$ is $\frac{1}{2}\Delta H$ for the reaction $Mg(Cp)_2(g) \rightarrow Mg(g) + 2Cp(g)$.

[c] $\bar{D}$ (ionic) is used to designate average ionic dissociation energy e.g. $\bar{D}$ for $[MgCp_2]^+$ is $\frac{1}{2}\Delta H$ for the reaction $[MgCp_2]^+ (g) \rightarrow Mg^+(g) + 2Cp(g)$. The positive charge is always on the reactant and on the metal product for the listed dissociations.

[d] $\Delta\bar{D}$ is the $\bar{D}$ (neutral) minus the $\bar{D}$ (ionic).

[e] These values are taken from the classic work of Friedman *et al*. (Ref. 11).

[f] $\bar{D}$ (neutral) calculated from thermochemical data from Ref. 4 is 3.2 eV and from Ref. 24 is 3.0 eV.

[g] $\bar{D}$ (neutral) calculated from thermochemical data from Ref. 10 is 2.7 eV and from Ref. 24 is 2.5 eV.

[h] $\bar{D}$ (neutral) calculated from thermochemical data from Refs. 15 and 19a is 2.2 eV and from Ref. 24 is 1.8 eV.

[i] $\bar{D}$ (neutral) calculated from thermochemical data from Ref. 19a is 3.1 eV.

[j] $\bar{D}$ (neutral) for $M(CO)_x$ (g) from thermochemical data from Ref. 12 is 1.2 eV (Cr); 1.6 eV (Mo); 1.8 eV (W); 1.2 eV (Fe) and 1.5 eV (Ni).

[k] See also Refs. 61a, 24b, 33a, and 24a for other data where the $\bar{D}$ (neutral) $\approx \bar{D}$ (ionic) for mono-metal carbonyls.

reactions leading to the formation of the metal atom and all other processes where the required radical ionization potentials were available. The radical I.P.'s were taken from one of the listed references, the values being reproduced as part of Table 3. In cases where both ionic and neutral values could be calculated the differences in the average dissociation energies are listed in the $\Delta\bar{D}$ column of the table. In some cases only one bond is ruptured and the average value is the specific bond dissociation value. The purpose of the tabulation is to show that the trends in the neutral bond dissociation energies are validly established by measurements of the ionic dissociation energies. The $\Delta\bar{D}$ values are small in the majority of cases and this leads to the further conclusion that the ionic values are fairly reliable for actual estimations of the neutral values. The large $\Delta\bar{D}$ for $FeCp_2$ and $CoCp_2$ can be explained on the basis of the unreliability of the radical ionization potentials used in the $\bar{D}$ (neutral) calculation. The values reported by Friedman *et al.*[11] for FeCp and CoCp were used because no direct measurements were available. The specific bond dissociation energy for rupture of the first Cp ring in metallocenes is usually large. The second Cp rupture generally requires much less energy[11,46]. Yet Friedman *et al.*[11] used the average of both M–Cp bond dissociation energies for the calculations of the MCp ionization potentials. Consequently the radical I.P.'s reported[11] are accurate only for those metallocenes where the two (M–Cp) bond energies are about equal which incidentally occurs only for $MgCp_2$ and $MnCp_2$. By assuming that the ionic and neutral bond energies are approximately equal the estimated I.P.'s of FeCp and CoCp are 7.1 and 6.2 eV compared to the 9.78 and 10.05 values given in Table 4 and used for $\bar{D}$ (neutral) calculations for $FeCp_2$ and $CoCp_2$.

Previous results for metal carbonyls where the average ionic and neutral dissociation energies are equal are included for comparison at the bottom of the table. This approximate equality of average bond dissociation energies should be valid for any organometallic compound whose ionization potential is near that of the metal atom. This suggestion is not meant to negate the effect of charge on bond energies but is meant to show that the effect is small when averaged over all the existing bonds and to the limits suggested by $\Delta\bar{D}$ values listed in the table. It is admitted that the normal A.P. (Appearance Potential) errors caused by unknown kinetic energy, electronic excitations and vibrationally excited states could affect either or both the ionic and neutral dissociation energies. Thus discrepancies between the neutral values listed here and the values calculated from calorimetric data (listed in the footnotes to Table 5) may be explained by invoking these probable errors in the A.P. measurements. Still, the probability also exists that the A.P. values do not contain any appreciable errors due to kinetic or excitation energies and yield true measurements of the "absolute" or "intrinsic"[24] bond dissociation energies of the complexes. If so, agreement with the calorimetrically derived values is not to be expected.

Real and apparent difficulties

In addition to the normal difficulties associated with using appearance potentials to derive bond dissociation energies, other very real problems exist.

Sometimes the fragment ion, whose appearance potential is necessary for calculation of the desired bond energy, is absent from the spectrum or of such low intensity that the measurement is impossible. Another problem is associated with multi-energy pathways for decomposition to the same fragment ion. This problem is particularly troublesome for complex molecules and is illustrated in Figures 5 and 14 where several different energy pathways (listed in Tables 6 and 7) for formation of MCp^+ and M^+ ions exist.

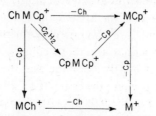

Figure 14 Multi-energy pathways for formation of MCp^+ and M^+ ions from ChMCp complexes where $M = V$ or Cr and $Ch = C_7H_7$ and $Cp = C_5H_5$
(see Table 7)

Table 6 Neutral products formed in the dissociation of $CpFeC_5H_4CH_2OH^+$ to MCp^+ and M^+ (see Figure 5)

Pathway	Ion	Proposed products Neutral(s)	Metastable obs.
1	$FeCp^+$	$C_5H_4CH_2OH$	yes
2	$FeCp^+$	$C_5H_4CH_2 + OH$	yes
3	Fe^+	$C_5H_4CH_2OH + Cp$	yes
4	Fe^+	$C_5H_4CH_2 + Cp + OH$	yes
5	Fe^+	$H_2O, C_5H_4CH_2 + C_5H_4$	yes
6	Fe^+	$H_2O, C_5H_3, C_5H_4CH_3$	no
7	Fe^+	multiple choice	no

Ion currents due to metastable transitions* were observed as indicated and are supplementary evidence that the listed pathways indeed exist. The multiple choice listing for pathway 7 in Table 6 and pathway 5 in Table 7 con-

---

* It should be stressed that the observations of metastable ion currents are positive proof that at least some of the ions decompose by the pathway established by the metastable observations. It does not imply that *all* ions must follow this pathway anymore than the absence of metastables imply that a suggested fragmentation scheme is impossible.

firm that these tabulations are illustrative rather than exhaustive for the listed ions. Müller[63] has reported data from which the energy requirements for pathways 1–3 in Table 7 may be calculated for the vanadium and chro-

Table 7 Neutral products formed in the dissociation of ChMCp (M = V or Cr) to MCp^+ and M^+ and the energy requirements for the various pathways (see Figure 14)

| Pathway | Proposed products | | Metastable[a] | Energy (eV)[a] | |
	Ion	Neutral(s)	obs.	V	Cr
1	MCp^+	Ch	yes	9.2	10.4
2	MCp^+	$Cp + C_2H_2$	yes	12.9	12.7
3	M^+	$Cp + Ch$	yes	13.8	12.2
4	M^+	$2(Cp) + C_2H_2$	yes	—	—
5	M^+	Multiple choice	no	—	—

[a] Taken from data reported in Ref. 63.

mium metallocenes. The energy differences here were large enough to allow deconvolution of the MCp^+ and M^+ ionization efficiency curves. With other complexes this may not always be the case and the long tail usually observed for the M^+ ion intensity near onset may be caused by unresolved energy pathways for formation of M^+ rather than the usual explanation based upon unknown kinetic and/or excitation energies.

A final problem, divorced from the measurements, has been alluded to in an earlier footnote. This is the tendency of some workers to estimate relative bond energies from observed ion intensities in mass spectra produced at high electron energies. While this approach has some validity when applied to ratios of like fragment ion currents from different but structurally similar molecules, it is without foundation when applied to a single complex molecule. For example, any relationship between the ratio of the CA^+ to AB^+ ion intensities from a (C–A–B) molecule and the ratio of the C–A to A–B bond dissociation energies is strictly fortuitous. Specifically, the ratio of the $FeCp^+$ to $FeC_5H_4XR^+$ ion currents is not indicative of the relative metal-ligand bond dissociation energies in the $CpFeC_5H_4XR$ molecule. If, however, the ratios are established for several different XR substituted complexes or different metal atom complexes, then the changes in these ratios are probably indicative of different *relative* bond energies. Even this approach is not theoretically sound and conclusions about relative bond energies derived solely from intensity measurements at electron energies of 50 or 70 eV should be supported with supplementary evidence obtained from ionization efficiency measurements or the results of other experiments.

Present utility and future developments

The usefulness of the measurements of the neutral bond dissociation energies does not require additional comment. The demonstrated ability of approximating the neutral values from ionic measurements for a large number of complexes should prove extremely useful. Serious attempts to deconvolute ionization efficiency curves in an attempt to obtain better ionic values for specific dissociations should also lead to more values for radical ionization potentials. Controlled thermal decomposition inside the ion chamber may prove a useful technique for measuring directly a radical I.P. in the presence of energetically interfering ion currents. More extensive use of ionization efficiency data, along with metastable observations and labelling experiments, to establish or confirm fragmentation schemes, should be forthcoming. It is hoped that bond energy measurements will be more extensively related to kinetic results for these complexes and used to propose reaction mechanisms. A reduction in reports relating ion intensity measurements to relative bond

Table 8   Comparison of the metal-ligand bond dissociation energies of like ions formed from different precursors

Fragment ion	Precursor	D (ionic)^a
CrBz	$CrBz_2$	2.0
CrBz	$BzCr(CO)_3$	2.6
CrCp	$CrCp_2$	1.8 (2.6)
CrCp	BzCrCp	4.6
CrCp	ChCrCp	1.8
MnCp	$MnCp_2$	2.5 (2.8)
MnCp	BzMnCp	4.7
MnCp	$CpMn(CO)_3$	3.9
VCp	VCp_2	1.8 (5.6)
VCp	ChVCp	4.6
VCp	$CpV(CO)_4$	5.2
CoCp	$CoCp_2$	0.7 (1.9)
CoCp	$CpCo(CO)_2$	5.2

^{a} Values taken from Table 5.

dissociation energies is expected as more laboratories become motivated and equipped to make ionization efficiency measurements. The effects of kinetic energy and electronic and vibrational excitations on the shapes of ionization efficiency curves may possibly be reassessed for metallocene complexes and even other complex molecules. Finally the formation of ionic and/or neutral products of unimolecular decompositions in their valency states as opposed to the ground state, may be shown to be prevalent.

The data in Table 8 illustrate the last development mentioned. It is admitted that errors in the appearance potential measurements caused by kinetic and excitation effects as well as possible errors due to multi-energy pathways may explain the metal-ligand bond dissociation energy discrepancies for the various MAr^+ ions formed from different parent molecules (precursors). Yet, it may be possible that these differences in bond dissociation energy are related to the formation of the various radical ions in different valence states. If so, the observation of different bond dissociation energies is related to the different valence states in the various molecules.

References

1. T.J.Kealy and P.L.Pauson, *Nature*, **168**, 1039 (1951).
2. S.A.Miller, J.A.Tebboth, and J.F.Tremaine, *J. Chem. Soc.*, **632** (1952).
3. G.Wilkinson, M.Rosenblum, M.C.Whiting, and R.B.Woodward, *J. Am. Chem. Soc.*, **74**, 2125 (1952).
4. F.A.Cotton and G.Wilkinson, *Ibid.*, 5764 (1952).
5. E.O.Fischer and W.Pfab, *Z. Naturforsch.*, **7B**, 377 (1952).
6. P.F.Eiland and R.Pepinsky, *J. Am. Chem. Soc.*, **74**, 4971 (1952).
7. J.D.Dunitz and L.E.Orgel, *Nature*, **171**, 121 (1953).
8. H.H.Jaffe, *J. Chem. Phys.*, **21**, 156 (1953).
*9. L.Friedman and A.P.Irsa, *J. Am. Chem. Soc.*, **75**, 5741 (1953).
10. G.Wilkinson, P.L.Pauson, and F.A.Cotton, *Ibid.*, **76**, 1970 (1954).
‡11. L.Friedman, A.P.Irsa, and G.Wilkinson, *Ibid.*, **77**, 3689 (1955).
12. F.A.Cotton, A.K.Fischer, and G.Wilkinson, *Ibid.*, **81**, 800 (1959).
*13. F.W.McLafferty, *Anal. Chem.*, **28**, 306 (1956).
14. M.Rosenblum and R.B.Woodward, *J. Am. Chem. Soc.*, **80**, 5443 (1958).
15. A.K.Fischer, F.A.Cotton, and G.Wilkinson, *J. Phys. Chem.*, **63**, 154 (1959).
‡16. A.G.Harrison, L.R.Honnen, H.J.Dauben, and F.P.Lossing, *J. Am. Chem. Soc.*, **82**, 5593 (1960).
‡17. R.W.Kiser, *Tables of Ionization Potentials*, TID-6142, U.S. Atomic Energy Commission, Office of Technical Information, June (1960).
17a. H.Zeiss, *Organometallic Chemistry*, Reinhold Publishing Corp., New York (1960).
18. E.A.Hill and J.H.Richards, *J. Am. Chem. Soc.*, **83**, 4216 (1961) and 3840 (1961).
18a. C.J.Ballhausen and J.P.Dahl, *Acta Chem. Scand.*, **15**, 1331 (1961).

19. D.J.Clancy and I.J.Spilners, *Anal. Chem.*, **34**, 1839 (1962).
19a. A.Reckziegel, *Dissertation*, Universitât München (1962).
*20. R.I.Reed and F.M.Tabrizi, *Appl. Spectry.*, **7**, 124 (1963).
21. M.D.Rausch, *J. Org. Chem.*, **28**, 3337 (1963).
†22. A.Mandelbaum and M.Cais, *Tetrahedron Letters*, 3847 (1964).
22a. C.D.Ritche and W.F.Sager, *Progress in Physical Organic Chemistry*, S.C.Cohen, A.Streitwieser, and R.W.Taft, editors, Interscience Publishers, New York (1964), p. 334.
‡23. E.Schumacher and R.Taubenest, *Helv. Chim. Acta*, **47**, 1525 (1964).
24. H.A.Skinner, *Adv. Organomet. Chem.*, **2**, 49 (1964).
‡24a. R.E.Winters and R.W.Kiser, *Inorg. Chem.*, **3**, 699 (1964).
‡24b. A.Foffani and S.Pignataro, *Z. Physik. Chem. Neue Folge*, **42**, 221 (1964).
*25. A.N.Nesmeyanov, T.V.Nikitina, and E.G.Perevalova, *Izv. Akad. Nauk SSSR, Ser. Khim.*, p. 197 (1964).
†26. A.N.Nesmeyanov, V.A.Dubovitskii, O.V.Nogina, and V.N.Bochkarev, *Dokl. Akad. Nauk SSSR*, **165**, 125 (1965).
‡27. A.G.Harrison, P.Haynes, S.McLean, and F.Meyer, *J. Am. Chem. Soc.*, **87**, 5099 (1965).
†28. N.Maoz, A.Mandelbaum, and M.Cais, *Tetrahedron Letters*, 2087 (1965).
*29. M.Cais, A.Modiano, and A.Raveh, *J. Am. Chem. Soc.*, **87**, 5607 (1965).
†30. I.J.Spilners and J.P. Pellegrini, *J. Org. Chem.* **30**, 3800 (1965).
*31. J.S.Shannon and J.M.Swan, *Chem. Comm.*, 33 (1965).
†32. A.F.Reid, J.S.Shannon, J.M.Swan, and P.C.Wailes, *Austral.J.Chem.*, **18**, 173 (1965).
‡33. R.E.Winters and R.W.Kiser, *J. Organomet. Chem.*, **4**, 190 (1965).
‡33a. R.E.Winters and R.W.Kiser, *Inorg. Chem.*, **4**, 157 (1965).
33b. M.Rosenblum, *Chemistry of the Iron Group Metallocenes*, Interscience Publishers, New York (1965).
*34. D.W.Slocum, R.Lewis, and G.J.Mains, *Chem. Ind.*, 2095 (1966).
†35. H.Egger and H.Falk, *Monatsh. Chem.*, **97**, 1590 (1966).
†36. H.Egger, *Ibid.*, 602 (1966).
†37. H.Egger and H.Falk, *Tetrahedron Letters*, 437 (1966).
38. M.Cais, *Organomet. Chem. Rev.* ,**1**, 435 (1966).
*39. M.Cais and N.Maoz, *J. Organomet. Chem.*, **5**, 370 (1966).
†40. H.Mechtler and K.Schlögl, *Monatsh. Chem.*, **97**, 754 (1966).
‡41. E.Schumacher and R.Taubenest, *Helv. Chim. Acta*, **49**, 1447 (1966).
‡42. R.G.Denning and R.A.D.Wentworth, *J. Am. Chem. Soc.*, **88**, 4619 (1966).
*43. W.E.Watts, *Ibid.*, 855 (1966).
44. W.E.Watts, *Organomet. Chem. Rev.*, **2**, 231 (1967).
45. W.E.Watts, *J. Organomet. Chem.*, **10**, 191 (1967).
‡46. J.Müller and L.D'Or, *Ibid.*, 313 (1967).
‡47. A.Foffani, S.Pignataro, G.Distefano, and G.Innorta, *J. Organomet. Chem.*, **7**, 473 (1967).
‡48. S.Pignataro and F.P.Lossing, *Ibid.*, **10**, 531 (1967).
*49. R.B.King and M.B.Bisnette, *Ibid.*, **8**, 287 (1967).
*50. E.O.Fischer, O.S.Mills, E.F.Paulus, and H.Wawersik, *Chem. Comm.*, 643 (1967).
*51. K.Schlögl and H.Soukup, *Tetrahedron Letters*, 1181 (1967).

†52. T.J.Katz and J.J.Mrowca, *J. Am. Chem. Soc.*, **89**, 1105 (1967).

*53. E.O.Fischer and M.W.Schmidt, *Angew. Chem.*, **6**, 93 (1967).

*54. E.O.Fischer and M.W.Schmidt, *Ibid.*, **79**, 99 (1967).

*55. R.Coutts and P.C.Wailes, *Inorg. Nucl. Chem. Letters*, **3**, 1 (1967).

*55a. M.I.Bruce, *Ibid.*, **3**, 157 (1967).

*56. B.D.James, R.K.Nanda, and M.G.H.Wallbridge, *Inorg. Chem.*, **6**, 1979 (1967).

†57. D.T.Roberts, W.F.Little, and M.M.Bursey, *J. Am. Chem. Soc.*, **89**, 4917 (1967).

†58. D.T.Roberts, W.F.Little, and M.M.Bursey, *Ibid.*, 6156 (1967).

‡59. D.R.Bidinosti and N.S.McIntyre, *Can. J. Chem.*, **45**, 641 (1967).

‡60. M.I.Bruce, *Adv. Organomet. Chem.*, **6**, 273 (1968).

60a. J.Lewis and B.F.G.Johnson, *Accounts Chem. Res.*, **1**, 245 (1968).

‡61a. G.A.Junk and H.J.Svec, *Z. Naturforsch.*, **23B**, 1 (1968).

62. S.Pignataro and F.P.Lossing, *J. Organomet. Chem.*, **11**, 571 (1968).

63. J.Müller and P.Göser, *Ibid.*, **12**, 163 (1968).

†63a. J.D.Hawthorne, M.J.May, and R.N.F.Simpson, *Ibid.*, 407 (1968).

†64. J.Müller and M.Herberhold, *Ibid.*, **13**, 399 (1968).

†65. W.E.Watts, Ferrocenophane Mass Spectra—see M.I.Bruce, *Adv. Organomet. Chem.*, **6**, 273 (1968).

*66. T.H.Barr and W.E.Watts, *Tetrahedron*, **24**, 3219 (1968).

‡67. P.Schissel, D.J.McAdoo, E.Hedaya, and D.W.McNeil, *J. Chem. Phys.*, **49**, 5061 (1968).

‡68. E.Hedaya, D.W.McNeil, P.Schissel, and D.J.McAdoo, *J. Am. Chem. Soc.*, **90**, 5284 (1968).

‡68a. S.Pignataro, A.Foffani, G.Innorta, and G.Distenfano, *Advances in Mass Spectrometry*, E.Kendrick ed., Institute of Petroleum, London (1968), p. 323.

†69. D.T.Roberts, W.F.Little, and M.M.Bursey, *J. Am. Chem. Soc.*, **90**, 973 (1968).

*70. F.Seel and V.Sperber, *J. Organomet. Chem.*, **14**, 405 (1968).

*71. R.B.King, *Organomet. Chem. Rev.*, **4**, 67 (1968).

*72. *Ibid.*, 3 (1968).

†73. M.Cais, M.S.Lupin, N.Maoz, and J.Sharvit, *J. Chem. Soc.*, A, 3086 (1968).

†74. M.S.Lupin and M.Cais, *J. Chem. Soc.*, A, 3095 (1968).

*75. E.O.Fischer, A.Reckziegel, J.Müller, and P.Göser, *J. Organomet. Chem.*, **11**, 13, (1968).

*76. E.O.Fischer and H.W.Wehner, *Ibid.*, 29 (1968).

†76a. M.I.Bruce, *Int. J. Mass Spectrom. and Ion Phys.*, **1**, 141 (1968).

†76b. M.I.Bruce, *Org. Mass Spectrom.*, **1**, 503 (1968).

†76c. *Ibid.*, 687 (1968).

†76d. M.I.Bruce and M.A.Thomas, *Ibid.*, 835 (1968).

†76e. M.L.H.Green, *Organometallic Compounds, Vol. 2—The Transition Elements*, Methuen and Co., Ltd., London (1968).

†77. M.Cais, M.S.Lupin and J.Sharvit, *Israel J. Chem.*, **7**, 73 (1969).

†78. R.B.King, *Appl. Spectry.*, **23**, 148 (1969).

79. J.G.Dillard and R.W.Kiser, *J. Organomet. Chem.*, **16**, 265 (1969).

†80. M.L.Anderson and L.R.Crisler, *Ibid.*, **17**, 345 (1969).

*81. P.M.Bruce, B.M.Kingston, M.F.Lappert, T.R.Spalding, and R.C.Srivastava, *J. Chem. Soc.*, **17**, 2106 (1969).

*82. D.E.Bublitz, *J. Organomet. Chem.*, **16**, 149 (1969).

‡83. J.Müller, *Chem. Ber.*, **102**, 152 (1969).

‡84. J.L.Thomas and R.G.Hayes, Private Communication (1969).

‡85. J.Müller and P.Göser, *Chem. Ber.*, **102**, 3314 (1969).

 85a. J.Müller and J.A.Conner, *Ibid.*, 1148 (1969).

‡86. G.E.Herberich and J.Müller, *J. Organomet. Chem.*, **16**, 111 (1969).

‡87. J.Müller, *Ibid.*, **18**, 321 (1969).

‡88. J.Müller and K.Fendrel, *Ibid.*, **19**, 123 (1969).

*89. M.I.Bruce, *Org. Mass Spectrom.*, **2**, 1037 (1969).

†90. M.I.Bruce, *Ibid.*, 997 (1969).

†91. M.I.Bruce, *Ibid.*, 63 (1969).

†92. G.A.Junk and H.J.Svec, unpublished results (1969).

†93. G.A.Junk and H.J.Svec, unpublished results (1969).

‡94. G.A.Junk and H.J.Svec, unpublished results (1969).

No sign—no mass spectral data.

* Little or no discussion of mass spectra.

† Fairly extensive discussion of mass spectra.

‡ Ionization efficiency curve interpretation used to establish I.P.'s and/or A.P.'s in addition to a discussion of mass spectral data.

Application of high-resolution mass spectrometry to complex mixtures

A. G. SHARKEY, JR.

U.S. Department of the Interior, Bureau of Mines
Pittsburgh Coal Research Center

INTRODUCTION

In the past 30 years the use of mass spectrometry has expanded from the laboratory of the physicist to the analytical chemist to the organic chemist. One reason for this expanded use of mass spectrometry in the field of organic chemistry is that there is a better understanding of the relationship between mass spectra and molecular structures resulting from studies by Meyerson[1], Djerassi[2], McLafferty[3], Reed[4], Biemann[5] and others. These studies have shown that in many instances the spectra produced by electronic impact can be interpreted in terms of chemical changes.[1]

A second development leading to the expanded use of mass spectrometry by the organic chemist has been the availability of high-resolution instruments. High-resolution mass spectrometry gives the chemist much additional information to use in structure elucidation. Atomic species in fragment ions as well as in the molecular ion can be deduced from precise mass values. A better understanding of decomposition mechanisms has been obtained from a knowledge of the species present in various fragment and rearrangement ions. Data processing schemes, either off-line or on-line, have made the high-resolution technique even more valuable for structural studies of pure compounds.[6]

While most of the high-resolution data reported to date has dealt with studies of pure compounds, several investigations of complex hydrocarbon mixtures have been made at the Pittsburgh Coal Research Center of the U.S. Bureau of Mines.[7] These investigations have involved primarily mixtures of polynuclear aromatic compounds derived from coal. A limited number of high-resolution studies of other mixtures have been reported by other laboratories including an investigation of petroleum wax by Reid,[8] an analysis of a petroleum stream by Johnson and Aczel,[9] a study of sulfur compounds in petroleum by Drushel and Sommers,[10] and a study of aromatics and saturates by Gallegos *et al.* using a type-analysis technique.[11] In each of the above investigations use was made of the capability of the instrument to determine precise masses and therefore distinguish among hydrocarbons with different C/H ratios or containing various heteroatoms.

First let us consider some of the limitations in the interpretation of spectra obtained on low or moderate resolution instruments. In dealing with mixtures we are limited by the number of structural types free of interference. Considering only the molecular ions for hydrocarbons, we are limited to 7 structural types before m/e values for alkyl derivatives overlap. A second type of spectral interference involves compound types such as alkylbenzenes and benzothiophenes which cannot be resolved with moderate resolution instruments. If we consider ions other than the parent ions, unique peaks such as those that arise with many oxygenated compounds (and also rearrangement ions) can in certain instances alleviate this situation. In other spectra the rearrangement ions simply add to the confusion. With a moderate resolution instrument we must resort to heavy isotopes to obtain clues to molecular formulas. Isotope groupings such as arise when alkyl halides are involved will assist in indicating the elements present.

High-resolution mass spectrometry is unique in providing a precise mass from which a molecular formula can be derived. The basis of the high-resolution technique is the ability of the instrument to resolve multiplets, having the same nominal mass but differing in precise mass, and to make precise mass assignments corresponding to the various combinations of elements. Instrumentation required for high-resolution mass spectrometry has been described in detail in the literature.[12] As indicated above there are two basic considerations: (1) The resolution required for separation and (2) the accuracy required of mass assignments. Both of these parameters vary according to the particular doublet or combination of species involved and the m/e value at which the peaks are observed. The most common and perhaps easiest way to consider resolving power for high-resolution measurements is in

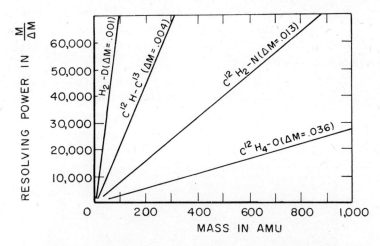

Figure 1 Required resolving power of some common doublets

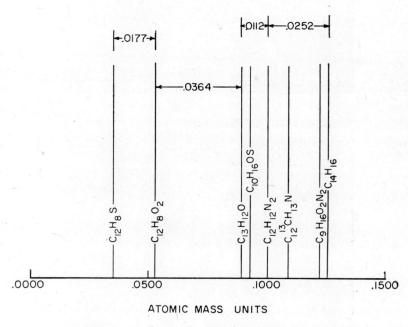

Figure 2 Relative position of several possible atomic combinations of *m/e* 184

terms of $(M/\Delta M)$—where (M) is the nominal mass at which the measurement is made and (ΔM) the separation of the doublet in atomic mass units.

The resolving power required to separate various doublets as a function of m/e is shown in Figure 1. In terms of atomic mass separation the ^{12}CH–^{13}C and H_2–D doublets are two of the most difficult to resolve. In dealing with complex mixtures the resolving power required to separate one doublet, as indicated in Figure 1, tells only part of the story. In many instances various combinations of atoms lead to complex multiplets as illustrated in Figure 2.

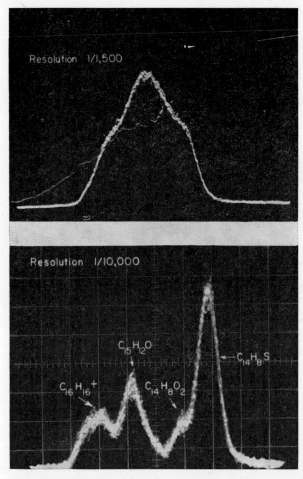

Figure 3   Nominal mass 208, coal-tar pitch

This added complexity, prevalent in the spectra of many mixtures, places a stringent requirement on the other parameter, namely accuracy of mass assignment. If one is dealing with only hydrocarbons and oxygenated species, for example, mass assignments with a precision of about 2 or 3 millimass units are acceptable. With the addition of other elements such as sulfur and nitrogen, mass assignments to within $\pm$ a few tenths of a millimass unit can be required. An oscilloscope trace of the m/e 208 region in the spectrum of coal-tar pitch is shown in Figure 3. The appearance of multiplets can be seen as the resolution is increased from approximately 1 in 1500 to 1 in 8000; four different species are evident.

Three types of data acquisition systems are (1) photographic plate, (2) magnetic tape, and (3) on-line computers. Each of these systems has certain advantages and disadvantages, and the details have been reported previously. The Nier peak matching system, a manual technique to obtain precise mass values for individual peaks, is inherently quite accurate but much too slow for obtaining the large quantity of data required. Computation of the elemental composition based on the precise mass value is generally handled by a computer, particularly where there are many determinations involved. Elimination of a repetitive feature such as $CH_2 = 14$ (for alkyl derivatives) or $C = 12$ shortens the calculation procedure considerably. One of the most popular means of displaying the data is use of an element map described by Biemann.[13] Data shown in the various columns can be correlated with fragments of the molecule for structural elucidation.

APPLICATION TO COMPLEX MIXTURES

In research on coal and other fuels one is generally dealing with mixtures and many times these mixtures are too complex to be resolved by any of the common separation and analytical techniques including combined gas chromatograph-mass spectrometry. Three types of investigations will be described to illustrate application of the high-resolution technique to complex mixtures.

1. Characterization studies—airborne particulate matter.

2. Compositional changes occurring with reaction—studies of weathered road tars.

3. Studies of reaction mechanisms—use of tagged species.

An extensive investigation has recently been made of a complex mixture of

Table 1 Summary of precise masses for molecular ions; organic material from airborne particulate matter

Mass, amu				Elemental composition				
Nominal	Theoretical	Measured	Δ, mmu	C	H	O	N	S
168	0939	0951	1.2	13	12			
170	0731	0746	1.5	12	10	1		
172	0888	0854	3.4	12	12	1		
178	0783	0775	0.8	14	10			
179	0735	0763	2.8	13	9		1	
180	0575	0589	1.4	13	8	1		
	0939	0915	2.4	14	12			
	1878	1864	1.4	13	24			
181	0640	0647	0.7	11	7		3	
182	0731	0733	0.2	13	10	1		
	1095	1057	3.8	14	14			
	2034	2021	1.3	13	26			
190	0783	0752	3.1	15	10			
192	0939	0935	0.4	15	12			
	1878	1887	0.9	14	24			
194	0732	0724	0.8	14	10	1		
195	0796	0777	1.9	12	9		3	
196	0524	0557	3.3	13	8	2		
	0888	0883	0.5	14	12	1		
	1252	1241	1.1	15	16			
	2191	2203	1.2	14	28			
202	0783	0791	0.8	16	10			
206	1096	1058	3.8	16	14			
208	0347	0353	0.6	14	8			1
	0888	0876	1.2	15	12	1		
216	0939	0910	2.9	17	12			
217	0891	0924	3.3	16	11		1	
220	0888	0893	0.5	16	12	1		
	1252	1232	2.0	17	16			
222	0503	0514	1.1	15	10			1
	1045	1037	0.8	16	14	1		

Table 1 *(cont.)*

Mass, amu				Elemental composition				
Nominal	Theoretical	Measured	Δ, mmu	C	H	O	N	S
224	0660	0661	0.1	15	12			1
226	0783	0783	0	18	10			
228	0939	0931	0.8	18	12			
229	0891	0924	3.3	17	11		1	
230	0732	0742	1.0	17	10	1		
	1095	1077	1.8	18	14			
232	0888	0860	2.8	17	12	1		
234	0503	0536	3.3	16	10			1
	1045	1002	4.3	17	14	1		
240	0939	0922	1.7	19	12			
242	0732	0718	1.4	18	10	1		
	1095	1068	2.7	19	14			
244	0888	0868	2.0	18	12	1		
245	1204	1212	0.8	18	15		1	
248	0660	0638	2.2	17	12			1
252	0939	0932	0.7	20	12			
276	0939	0943	0.4	22	12			
300	0939	0925	1.4	24	12			

polynuclear aromatic compounds derived from airborne particulate matter.[14] Results of this investigation will be used to illustrate application of the high-resolution technique to the characterization of complex mixtures. Results of studies of altered samples and use of tagged species have been described in detail in previous reports and will be mentioned only briefly.[15,16,17]

Organic contaminants in airborne particulate matter

Although a wide variety of organic contaminants, including chemical carcinogens, have been identified in air from various localities little use has been made of modern instrumental techniques such as high-resolution mass spectrometry to further these studies. Sawicki and co-workers have made extensive studies of organic compounds found in urban atmospheres.[18] Because separation and identification of the individual components is difficult, the concentrations of 3,4-benzopyrene and one or two other polynuclear aromatics have been used as a basis for comparing organic pollutants in atmospheres.[19] High-resolution mass spectrometry can assist in determining the presence of various contaminants containing heteroatoms, including polynuclear carbonyl compounds already under suspicion as possible carcinogens. While the specific structural types cannot be identified, carbon number distribution data for the various alkyl derivatives can be determined. Quantitative analyses for alkyl derivatives can be made (without standards for the derivatives) using the low ionizing voltage technique.

Samples of airborne particulate matter collected in the Oakland area of Pittsburgh were supplied by the Graduate School of Public Health, University of Pittsburgh. Two milligram samples of solid particulate matter were introduced in the inlet system of the mass spectrometer and heated to approximately 300°C; about 12 percent of the material was vaporized at this

Table 2 High-resolution mass spectrometric data for several organic pollutants in airborne particulate matter

Mass, amu				Elemental composition				Examples of possible compounds[b]
Nominal	Theoretical	Measured[a]	Δ	C	H	O	N	
C, H								
178	0.0783	0.0775	0.0008	14	10			Anthracene
202	0.0783	0.0774	0.0009	16	10			Pyrene
228	0.0938	0.0930	0.0008	18	12			Benzanthracene
252	0.0938	0.0932	0.0006	20	12			Benzopyrene
276	0.0938	0.0943	0.0005	22	12			Benzo[ghi]perylene
300	0.0938	0.0925	0.0013	24	12			Coronene
C, H, N								
229	0.0981	0.0944	0.0037	17	11		1	Benzacridine
C, H, O								
180	0.0575	0.0589	0.0014	13	8	1		Phenalen-1-one
230	0.0732	0.0742	0.0010	17	10	1		Benzanthracen-7-one
C, H, O$_2$								
196	0.0524	0.0557	0.0033	13	8	2		Xanthen-9-one

[a] Exact fractional mass for elemental composition shown.
[b] Previously identified air pollutants; see Ref. 17.

temperature. High-resolution mass spectra were obtained at a resolution of approximately 1 part in 15,000. Precise masses used in determining the elemental compositions were, with a few exceptions, assigned with accuracies of 1 to 3 millimass units.

Precise mass data were obtained for approximately 750 peaks in the mass 76 to 266 range. Spectra obtained at low-ionizing voltage were used in conjunction with reports of previously identified pollutants in determining molecular ions (Table 1). Components having up to 2 oxygen and 3 nitrogen atoms were indicated. The five most intense peaks in the mass spectrum were m/e 178, 202, 228, 252, and 276. Elemental compositions derived from the precise masses and examples of known pollutants corresponding to these formulas are shown in Table 2. Also shown in Table 2 are examples of known pollutants containing CHN and CHO.

Data given in Table 3 illustrate the ability of the instrument to detect organic compounds having the same nominal molecular weight but different precise masses resulting from the presence of heteroatoms. For example, compounds having the formulas $C_{13}H_{12}$ and $C_{12}H_8O$ were detected at mass 168. The detection of alkyl derivatives is particularly important as it has been shown that in some instances alkyl derivatives are more carcinogenic than unsubstituted aromatic ring structures.[20] Alkyl derivatives containing up to 4 carbon atoms are indicated for 7 of the ring systems shown in Table 3.

The components at m/e 230 are of particular interest as it has recently been shown that benzanthracen-7-one is in certain tests many times more carcinogenic than well-known carcinogens such as benzopyrene. Resolution of the doublet indicated at mass 230 yielded one component with a formula corresponding to benzanthracen-7-one, $C_{17}H_{10}O$. The formula of the other component corresponded to a previously identified air pollutant, dimethylpyrene, $C_{18}H_{14}$.

The general features of the spectrum, and in particular the mass range of the structural types containing heteroatoms, can be readily determined by tabulating the data obtained with 70-volt electrons in a form similar to the element map used for elucidating the structure of pure compounds.[13] Data tabulated in this manner for a portion of the high-resolution spectrum of the volatiles from airborne particulate matter are shown in Table 4. The designations used are the same as for the element map described by Biemann with the type and number of heteroatoms shown in the headings. The number of carbon and hydrogen atoms are designated in the blocks at each m/e value.

Table 3 Partial high-resolution mass spectrum of organic contaminants in airborne particulate matter

Nominal mass	Relative intensity	Elemental composition				Examples of possible compounds[a]
		C	H	O	N	
154	6.9	12	10			Acenaphthene[b]
168	6.5	13	12			Methylacenaphthene
	4.0	12	8	1		Dibenzofuran[b]
182	6.2	14	14			Dimethylacenaphthene
	2.7	13	10	1		Benzophenone
196	5.5	15	16			Trimethylacenaphthene
	2.5	13	8	2		Xanthen-9-one[b]
167	12.8	12	9		1	Carbazole[b]
181	5.3	13	11		1	Methylcarbazole
166	12.0	13	10			Fluorene[b]
180	9.3	14	12			Methylfluorene
	2.6	13	8	1		Phenalen-1-one[b]
194	5.0	15	14			Dimethylfluorene
	4.5	14	10	1		Phenanthrol
178	22.0	14	10			Phenanthrene; anthracene[b]
192	12.0	15	12			Methylphenanthrene[b]
206	8.6	16	14			Dimethylphenanthrene[b]
202	46.6	16	10			Pyrene; fluoranthene[b]
216	12.1	17	12			Benzofluorene[b]
230	9.0	18	14			Dimethylpyrene[b]
	6.0	17	10	1		Benzanthracen-7-one[b]

[a] All isomeric variants are possible.
[b] Previously identified air pollutants; see Ref. 17.

In certain instances, multiplets derived from ions with different degrees of saturation are indicated.

The purpose of a tabulation of this type is to indicate the elemental composition of the fragment and molecular ions, and provide clues as to how the heteroatoms are bound in at least the major components in the mixture. The highest m/e indicated (for peaks above the trace level) was m/e 266. The three most heavily populated columns indicated in Table 4 are those for the CH, CHN, and CHO_2N_2 groupings. The C/H ratio in all three columns indicates the presence of aliphatic and aromatic moieties. The complete tabulation

Table 4   Composition of ions from volatiles in airborne particulate matter — (m/e 240–266)

Nominal mass	CH	CHN	CHO₂N₂	CHO	CHS	CHN₂	CHOS	CHN₃	CHON
240	17/36 19/12 18/24	16/34 17/22	13/24 14/12						
241	19/13 18/25	17/23 18/11	13/25 14/13						
242	19/14 18/26	17/24 18/12	14/14				15/14		
243	18/27	17/25 18/13	13/27 14/15						
244	19/16	17/26 18/14	13/28 14/16	18/12				16/10	
245		18/15	13/29	18/13					
246	20/6 19/18 18/30	17/28 18/16	13/30 14/18	18/14	17/10				
247	20/7 19/19 18/31	17/29 18/17	14/19	17/27	17/11				17/11
248	20/8 19/20 18/32	17/30 18/18	13/32		17/12				17/12
249	20/9 18/33	17/31 18/19							17/13
250	20/10 18/34	17/32 18/20	13/34						
251	20/11 19/23 18/35	19/9 17/33 18/21	15/11 14/23						
252	20/12 19/24 18/36	17/34 18/22	15/12 14/24						
253	18/37	19/11 17/35							
254	20/14	19/12	15/15 14/27			18/10	15/26		
255	20/15 19/27	19/13 18/25	15/22	19/11				17/ 9	
262	20/22 19/34	19/20 18/32	16/14						
266	20/26 19/38 21/14	19/24 20/12 18/36							

shows that ions composed of CH occur over the entire mass range investigated, m/e 76 to 266. Ions composed of CHN first appear with 14 carbon atoms at mass 194, indicating at least one nitrogen is bound into a stable ring system. Ions composed of CHO_2N_2 are indicated first at m/e 112, corresponding to $C_4H_4O_2N_2$. Based on the C/H ratio, the two oxygen and nitrogen atoms are associated with an unsaturated system. Ions composed of CHO, CHS, CHN_2, CHOS, CHN_3, and CHON were detected at only a few m/e values.

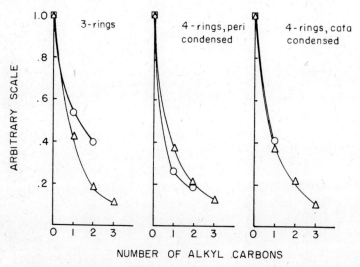

△ Coal-tar pitch
○ Airborne particulates (Pgh)

Figure 4   Relative concentrations of alkyl derivatives of polynuclear aromatic hydrocarbons found in airborne particulate matter and coal-tar pitch

Solid fuel combustion is a well-known source of airborne contaminants and it is of interest to interpret these results in terms of high-temperature combustion products from coal. The distributions of alkyl derivatives for 3-ring (mol.wt. 178), 4-ring peri- (mol.wt. 202), and 4-ring cata-condensed (mol.wt. 228) structures in coal-tar pitch and airborne particulate matter are shown in Figure 4. Carbon number distribution data are very similar for these particular polynuclear aromatics from the two sources. It should be emphasized that the technique determines only the number of alkyl carbons, the type of alkyl substituent(s) and position(s) on the ring(s) are not determined.

Composition changes occurring with reaction

Use of the high-resolution mass spectrometric technique to study compositional changes occurring in samples exposed to various atmospheres is illustrated by the investigation of a road tar. Mass spectra were obtained of the unprecipitated fractions from the unexposed tar and the tar after 12 and 24 months of weathering under road conditions. Molecular formulas and relative intensities were obtained for the multiplets appearing at the various nominal masses. Percentage of total ionization based on the summation of intensities for ions containing CH, CHO, and CHS, was used as the basis for comparison (Table 5). At molecular weights less than 200, decreases in

Table 5   Weathered road tars

Exposure time, months		None (fresh)	12	24	12	24
Atomic species	Molecular weight range	Total	ionization	percent	Percent change compared to fresh road tar	
CH	<200	11.6	9.0	7.4	−22.4	−36.2
	200–320	14.3	15.7	16.6	+9.8	+16.1
CHO	<200	1.9	1.5	1.2	−21.1	−36.8
	200–320	3.5	4.2	4.3	+20.0	+22.9
CHS	<200	0.61	0.58	0.41	−4.9	−32.8
	200–320	0.95	1.04	1.12	+9.5	+17.9
Total sample	<200	16.4	12.1	10.0	−26.2	−39.0
	200–320	19.5	21.7	23.0	+11.3	+17.9

ionization attributable to polynuclear aromatics, oxygenated components, and the total sample are similar. The increase in percentage of total ionization of the oxygen-containing material between masses 200 and 320, relative to the original road tar, is more than double that of polynuclear aromatic ring systems after 12 months of exposure and 50 percent higher after 24 months of exposure. Data shown in Table 5 represent a condensation of several hundreds of pieces of data obtained by high-resolution mass spectrometry. Information concerning the formulas of components showing changes during weathering gives a new insight into the process. It should be emphasized that these data were obtained on essentially total samples with little separation involved.

Studies of reaction mechanisms

Fu and Blaustein have investigated the reaction of coal and of graphite in a microwave discharge with H_2O or D_2O present.[21] They found that gasification of hva bituminous coal in an argon discharge is enhanced in the presence of H_2O. D_2O was used in the place of H_2O in a set of similar experiments to determine if hydrogen present in the coal and/or H species produced from H_2O in the discharge contributed to hydrocarbon formation.

Experiments with coal and D_2O in a microwave discharge indicated that the gaseous products contain substantial amounts of deuterated hydrocarbons. Analysis of this complex product was performed with a high-resolution mass spectrometer at a resolution of about 1 part in 20,000. Thus, high-resolution is required even in the C_2 hydrocarbon region as the H_2–D doublet is separated by only 1.5 millimass units. The m/e 26, 27, and 28 region representing several of the major products is shown in Table 6. The product contained undeuterated, partially deuterated, as well as completely deuterated acetylene. It appeared that hydrogen was derived from both the coal and the H_2O. HCN, another product, was easily detectable at mass 27, even in this complex mixture.

Table 6 Microwave discharge reaction products Coal-D_2O (Resolution 1/20,000)

Nominal mass	Measured mass	Actual mass	Δamu	Structure
26	26.0026	26.0031	−0.0005	CN^+ (fragment ion)
	26.0136	26.0141	−0.0005	C_2D^+ (fragment ion)
	26.0154	26.0156	−0.0002	C_2H_2
27	27.0100	27.0109	−0.0009	HCN
	27.0227	27.0219	+0.0008	C_2DH
28	27.9929	27.9949	−0.0020	CO
	28.0041	28.0061	−0.0020	N_2
	28.0273	28.0282	−0.0009	C_2D_2

CONCLUSIONS

The above examples illustrate the applicability of high-resolution mass spectrometry to complex mixtures of polynuclear aromatic hydrocarbons and reactions involving tagged species. Much information can be derived from studies of essentially total samples with only minimal separation. Information concerning alkyl derivatives such as carbon number distribution data could be of prime importance in studies of carcinogenic pollutants.

Acknowledgments

The contributors of R.A. Friedel, Janet L. Shultz, Theodore Kessler, and Joseph Malli, of the U.S. Bureau of Mines, Pittsburgh Coal Research Center, to the above mass spectrometric studies is gratefully acknowledged.

References

1. Meyerson, Seymour, *Record of Chemical Progress*, **26**, No. 4, 257–267, Dec. 1965.
2. Budzikiewicz, H., Djerassi, C., and Williams, D.H., *Interpretation of Mass Spectra of Organic Compounds*, Holden-Day, Inc., San Francisco, 1964.
3. McLafferty, F.W., Ed., *Mass Spetrometry of Organic Ions*, Academic Press, New York, 1963.
4. Reed, R.I., *Applications of Mass Spectrometry to Organic Chemistry*, Academic Press, New York, 1966.
5. Biemann, K., *Mass Spectrometry—Organic Chemical Applications*, McGraw-Hill Book Co., Inc., New York, 1962.
6. Burlingame, A.L., D.A.Smith, and R.W.Olsen, *Anal. Chem.*, **40**, 13–19, Jan. 1968.
7. Sharkey, A.G., Jr., J.L.Shultz, T.Kessler, and R.A.Friedel, Ch. in *Spectrometry of Fuels*, R.A.Friedel, Ed. (in Press).
8. Reid, W.K., *Anal. Chem.*, **38**, No. 3, 445–449, 1966.
9. Johnson, B.H., and Thomas Aczel, *Anal. Chem.*, **39**, No. 6, 682–685, 1967.
10. Drushel, Harry V., and A.L.Sommers, *Anal. Chem.*, **39**, No. 14, 1819–1829, Dec. 1967.
11. Gallegos, E.J., J.W.Green, L.P.Lindeman, R.L.LeTourneau, and R.M.Treater, *Anal. Chem.*, **39**, No. 14, 1833–1838, 1967.
12. Beynon, J.H., *Mass Spectrometry and Its Applications to Organic Chemistry*, Elsevier Pub. Co., Amsterdam, 1960.
13. Biemann, K., and P.V.Fennessey, *Chemia*, **21**, 226–235, June 1967.
14. Sharkey, A.G., Jr., J.L.Shultz, T.Kessler, and R.A.Friedel, *Research/Development*, **20**, No. 9, 30–32, Sept. 1969.
15. Sharkey, A.G., Jr., J.L.Shultz, T.Kessler, and R.A.Friedel, Preprints, *7th Intern'l Conf. on Coal Science*, Prague, Czechoslovakia, June 10–14, 1968.
16. Shultz, J.L., T.Kessler, R.A.Friedel, and A.G.Sharkey, Jr., *Proc. ASTM Com. E-14 Meeting*, May 12–17, 1968, pp. 359–361.
17. Kessler, T., and A.G.Sharkey, Jr., *Spectroscopy Letters*, **1**, No. 4, 177–180, 1968.
18. Sawicki, E., T.R.Hauser, W.C.Elbert, F.T.Fox, and J.E.Meeker, *Am. Indus. Hygiene Asso. Jour.*, **23**, 137–144, 1962.
19. Stanley, T.W., M.J.Morgan, and E.M.Grisby, *Environ. Sci. and Tech.*, **2**, No. 9, 699–702, 1968.
20. Sawicki, E., *Arch. Environ. Health*, **14**, 46–53, 1967.
21. Fu, Y.C., and B.D.Blaustein, *Chem. and Ind.*, 1257–1258, 1967.

Photoionization and photoelectron spectroscopy

JACQUES E. COLLIN

Laboratoire d'Etude des Etats Ionisés, Institut de Chimie, Université de Liège

I INTRODUCTION

Most ionization processes are produced by electron or photon impact and the interpretation of experimental data, particularly the ionization efficiency curves, has been discussed many times in review papers[1,2]. This work will rather emphasize developments in the recent years in the restricted, but of fundamental importance, area of photon impact ionization. Due to the great amount of published material, this will not be a review paper; instead we shall call attention upon the important fields of present active research.

Practically, most discussions are based on the consideration of either electron or photon impact ionization efficiency curves, or of photoelectron kinetic energy analysis and it may be useful to recall briefly the simple threshold laws[1,2] which are generally taken as a satisfactory first approximation basis for the discussion of the data. It is considered that photon impact excitation follows a resonance threshold law; whereas photoionization obeys a step function law; for electron impact, excitation is ruled by a step function law and ionization by a linear function of the excess energy of the impinging particle*. Ideally therefore the experimental curves as well as their first

---

* This is only a crude approximation generally valid at most for the very first eV above threshold. In photoionization it is fairly true for rare gases except in the autoionization region between the doublet components[3], and for polyatomic molecules where no pre-ionization occurs such as ethylene[4] but not for instance for CO_2 where the cross section rapidly decreases[5]. In electron impact, see for instance the work of Brion *et al.*[33] recently published.

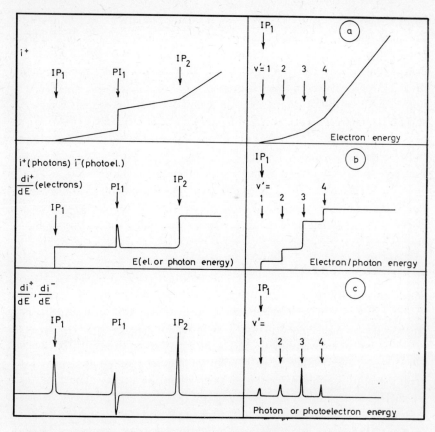

Figure 1   Idealized threshold laws for electron or photon impact preionization (PI) or ionization (IP); IP_1, IP_2 represent threshold energies for various electronic ionic states; PI_1 represents the threshold energy for a preionization process. Right side curves represent the case where a vibrational structure is obtained; a) refers to the direct electron impact curve; b) the direct photon or photoelectron curve or the first derivative of the electron impact curve; c) the first derivative of the photon impact or photoelectron curve

derivative as a function of energy should look as shown in Figure 1 and should allow a distinction between ionization and preionization processes.

All these processes are normally ruled by the Franck–Condon principle and the most probable transition to a given ionic state may not correspond to the lowest energy one, corresponding to the true or so-called "adiabatic" ionization potential (IP). In such a case, the measured IP is called the "vertical" IP. This obviously will depend on the relative positions and shapes of the potential energy curves or surfaces for both the molecule and the ion. As a consequence, if a transition occurs to an attractive ionic state in its stable energy region, i.e. below its dissociation limit, various vibrationally excited ions may appear and a corresponding vibrational fine structure may be observed. In addition, the relative intensities in the vibrational transitions should correspond to the transition probabilities, and these in turn are, in first approximation, proportional to the calculated Franck–Condon factors[2]. Therefore, additional information on the nature of the ionic states produced may be deduced from the efficiency curve shape.

The comparison of the data obtained by photoionization with those obtained either by electron impact or by UV spectroscopy has been discussed elsewhere[1,6]. Although UV absorption may give very accurate data on ionization energies, it may be very difficult to interpret, owing to the complexity or diffuseness of the spectra. The interpretation is bound to the elucidation and classification of absorption lines in Rydberg series; these may then be discussed, notably in terms of quantum defects. This has been recently done extensively by Lindholm and co-workers[7] using UV and electron energy loss data such as obtained by Lassettre and co-workers[8].

On the other hand one should not consider that electron impact data are obsolete, only because selection rules are often less drastic, particularly for spin-forbidden transitions, for low energy electron impact so that forbidden transitions may then be detected and yield additional information. This has recently been discussed[9] and is currently investigated in electron energy loss experiments such as those of Lassettre *et al.*[8,10] or using the trapped-electron technique[11] or the scavenger-SF_6 method[12,13]. Moreover very important improvements have been achieved in recent years in obtaining monoenergetic electron beams, now of the order of 10–20 mV resolution, so that vibrational and rotational excitation of H_2 for instance has been observed[14].

II EXPERIMENTAL TECHNIQUES IN PHOTOIONIZATION AND PHOTOELECTRON SPECTROSCOPY

As mentioned above, the possible photon impact processes are the following:

$$M_{E',v',j'} + h\nu \rightarrow M^+_{E'_i+,v',j'} + e \quad \text{direct ionization} \tag{1}$$

$$\left\lbrace \begin{array}{ll} M_{E'',v'',j''} + h\nu \rightarrow M^*_{E',v',j'} & \text{superexcitation} \\ M^*_{E',v',j'} \quad\quad\quad \rightarrow M^+_{E'_i+,v',j'} & \text{preionization} \end{array} \right\rbrace \quad \begin{array}{l} (2a) \\ (2b) \end{array}$$

In both (1) and (2) the excited species appeared may also photodissociate either directly or by predissociation, yielding neutral or charged fragments. E, v, j, represent electronic, vibrational and rotational states of the molecule, excited molecule or ion. Since we deal with photon absorption processes, forbidden transitions will be highly improbable and a direct comparison with spectroscopy should be feasible.

From the experimental point of view, the main problems concern the light source, the collision chamber and the detection and analysis of the products of reaction.

1   Light sources

The problem of UV light sources is discussed in the book of Samson[15] and the reader is referred to it for detailed information.

A continuous UV light source is generally desirable, particularly if one is interested in getting photoionization curves. However it is easier to produce a high intensity line spectrum, particularly in resonance discharge tubes, so that such sources will be convenient for photoelectron spectroscopy at fixed wavelength. Continuous light sources may be obtained by a DC glow discharge in hydrogen, the useful spectral range being between 1600 and 5000 Å. Below 1600 Å, the line spectrum is by far the more intense, although the relative intensities of the continuum and line spectrum vary with pressure and lamp power conditions. With a capillary lamp and repetitive flash discharge, the Lyman continuum may be obtained down to 900 Å but is not very suitable for photoionization studies. Good continua were found in rare gases, including the Hopfield helium continuum. The useful spectral ranges are given in Table 1. They are obtained by high voltage repetitive spark discharge or by microwave discharge.

Table 1   Rare gas continua (from Samson[15])

Gas	Range (Å)	Pressure (mm)	Intensity (photons/sec)
He	580–1100	40–55	$\sim 10^8$
Ne	740–1000	> 60	—
Ar	1050–1550	150–250	$\sim 10^8$
Kr	1250–1800	> 200	$\sim 10^7$
Xe	1480–2000	> 200	$\sim 10^7$

Very short wavelength radiation from the synchrotron is another possible source and has indeed been used in the 100–600 Å range by Madden and Codling[16] who obtained in this way the absorption spectra of the rare gases below 600 Å[17].

A convenient line source is obtained with the cold cathode discharge in rare gases. Fairly monochromatic light of high intensity may be obtained in the 8–22 eV energy range, a range very useful for the study of photoionization. These sources are due to the first resonance lines of the neutral rare gases. The Lyman α line of hydrogen is also convenient. Table 2 summarizes the most interesting resonance lines.

Table 2   First resonance lines

Atom	Wavelength (Å)	Energy (eV)	Atom	Wavelength (Å)	Energy (eV)
He	584.3	21.21	Kr	1164.9	10.64
Ne	735.9	16.85		1235.8	10.03
	743.7	16.67	Xe	1295.6	9.57
Ar	1048.2	11.83		1469.6	8.44
	1066.7	11.62	H		
			(Lym α)	1215.2	10.21

To obtain a photoionization efficiency curve requires a variable wavelength light source. Therefore, the sources mentioned above must be followed by a dispersion system, in most cases a Seya–Namioka $-70°$ monochromator[18,19]. Commercial UV monochromators of that type are now available.

2   Methods of analysis

2.1 *Photoionization*

Photoionization curves are obtained by plotting the positive ion current as a function of wavelength, using a monochromator. Various authors have published such curves since the original work of Watanabe[20], Lossing and Tanaka[21], Hurzeler *et al.*[22] and Weissler *et al.*[23]. Particularly active in the field are Inghram and collaborators[24], Chupka, Berkowitz and co-workers[25], Dibeler and collaborators[26], Momigny *et al.*[27], Cook *et al.*[28] The curves were obtained either with or without mass analysis of the product ions. In the best cases[25], the resolving power was as high as 0.8 Å in the 1000 Å range. Ion multipliers or counting techniques have been used.

As we shall not be interested here in absolute cross section measurements, the interested reader will consult the books on ionization phenomena and particularly the review paper by Samson[3] and the book of Marr[29].

2.2 Photoelectron spectroscopy

In recent years a great effort has been made towards the study of photoionization by means of the kinetic energy analysis of the ejected photoelectrons, based on the simple energy relation where KE

$$KE\,(e^-) = h\nu - IP_{E_i^+, v', j'} \tag{3}$$

is the kinetic energy of the photoelectron, $h\nu$ is the energy of the photon beam and IP refers to the energy required to produce the ion in its i-th electronic state with the vibrational v' and rotational j' quantum numbers. Equation (3) is a very good approximation because according to the law of conservation of momentum practically all the total photon excess energy is used as kinetic energy of the photoelectron. Other sources of error have been discussed by Turner[30,31] and include the Doppler effect, the velocity of the target molecule and the broadness of the energy states having very short life-times as a consequence of the uncertainty principle. All these errors may be considered as negligible (<0.001 eV) in the presently available resolution conditions. Use of equation (3) may lead to the determination of very accurate IPs and vibrational ionic levels.

The simplicity of the method is a great advantage from both experimental and interpretational standpoints. In particular, if photoelectron spectroscopy is performed at fixed wavelength, only direct ionization processes (reaction (1) above) occur in principle and will not be obscured by preionization as the latter only takes place if the incident light energy exactly matches the energy of the preionization transition (resonance process), a case likely to be uncommon*.

On the other hand, as photoionization cross-sections are by no means a constant function of energy above threshold but rather tend to decrease, it may be desirable to operate with various wavelengths so as to ensure the appearance of ionization processes which may not be observed because of too low cross-sections and in order to obtain information on the variation of cross-section with energy. An example of such an application has been recently given by Natalis, Collin and Momigny[32] for benzene and fluoro-

---

* It will be shown later this simplification is, however, quite often not observed and that preionization is required to explain the experimental results.

benzene. Various monochromatic light sources or a monochromator may be used.

If one replaces the UV beam by a soft X-ray beam, ionization of electrons from inner shells may occur and photoelectrons of high energies are ejected. This is the basis of the new and powerful ESCA method proposed by Seigbahn and collaborators[34]. We shall consider it later in this paper.

Many ways of performing the electron kinetic energy analysis have been used, either based on a retarding field technique or on energy selection. Retarding field methods were first used in a slotted-cylindrical grid system by Vilesov *et al.*[35] and in a cylindrical Lozier apparatus by Scheen whereas Turner *et al.*[36] used a cylindrical grid system. A high resolution slotted-grid system has been recently developed by Price and co-workers[37,38] and a spherical-grid system was constructed by Frost and co-workers[39]. A retarding field electron lens was used by Brehm and von Puttkamer[40] with a resolving power of 13 mV*.

Various energy analyzers have also been applied: the electrostatic-127° type[31] or 180°-magnetic type[41], and the hemispherical-180° electrostatic type[42]. Recently a parallel-plate high resolution instrument was developed by Eland and Danby[43].

It is often desirable to measure accurate relative transition cross sections, either for a knowledge of relative ionic electronic transition probabilities or for a determination of vibrational transition probabilities for comparison with the theoretically calculated Franck–Condon factors[2]. It is therefore important to have a 100% collection efficiency. Obviously some geometrical arrangements may be preferred for such measurements, if there is an angular non-isotropic photoelectron distribution. From that standpoint, spherical systems seem particularly well suited. Theoretical papers, notably by Bethe[44] and Cooper and Zare[45] have shown that the angular distribution of photoelectrons ejected from a given level follows the law

$$i^- \div A + B \cos^2 \theta \qquad (4)$$

where i^- represents the number of photoelectrons ejected per unit solid angle by plane-polarized light, θ is the angle of ejection relative to the direction of the electric vector, i.e. perpendicular to the optical path, and A and B are coefficients depending on the energy and angular momentum of the photoelectrons.

---

* This instrument has now a resolution of 9 mV (H. Ehrhardt, private communication, 1969).

Application of equation (4) allows the angular distribution of electrons to be used as an indication of the nature of the orbital from which the electrons are ejected.

Angular distributions of photoelectrons have been determined experimentally by various authors in different geometrical arrangements: cylindrical system by Vilesov and co-workers[46], variable angle collection by Ehrhardt *et al.*[47,48], spherical systems by Frost *et al.*[49], whereas Samson has shown[50] analytically that observations at an angle of 54°44' in a spherical device should provide the average number of electrons ejected per unit solid angle, independent of any angular distribution, and should thus be of interest for the determination of relative cross sections.

Whether working at fixed wavelength or with a variable light source, it would be very useful to collect and analyze simultaneously the two charged products of reaction, positive ions and electrons. This idea has recently been applied in a very nice experiment by Brehm and von Puttkamer[40] who measured simultaneously by a coincidence technique both positive ions and photoelectron currents produced in the photoionization of CH_4 and NH_3.

III   THE STUDY OF IONIZATION PROCESSES BY PHOTO-IONIZATION AND PHOTOELECTRON SPECTROSCOPY

1   Introduction

Being capable of a high resolving power, photoionization and photoelectron spectroscopy are directly comparable to the more conventional UV and vibrational spectroscopy. The amount of information which is obtained is considerable and we shall show with a few examples what the possibilities are and what problems are encountered. The former include the determination of IP's and ionization cross sections, of vibrational levels in various electronic ionic states, of Rydberg series, of preionization and predissociation, of internal conversion, and the application to the elucidation of structural problems in a manner similar to IR and NMR spectroscopy.

2   The photoionization of H_2. Vibrational and rotational ionic states

The hydrogen molecule has been much studied and represents a very good example of the power of photoelectron spectroscopy, as compared to photoionization. The photoionization curve of hydrogen shows many preionization peaks[51-55] which have been discussed by Dibeler *et al.*[52] and by Chupka and Berkowitz[53] in relation with the spectroscopic work of Beutler and Junger[56] and Monfils[57]. The highest resolution was obtained by Chupka and Berkowitz (0.008 eV). The complicated preionization structure observed neither allows a clear separation of the vibrational levels of H_2^+ nor *a fortiori* a measurement of the vibrational transition probabilities. In contrast with the photoionization curve, the photoelectron spectrum obtained with He 584 (21.21 eV) i.e. in an energy region where no preionization occurs, is completely different and shows no preionization structure at all. The vibrational structure appears then very clearly[34,39,41,42,58], particularly in high resolution studies. From that point of view the best results seem to be those published by Siegbahn and collaborators[34] where a very nice vibrational structure is observed up to $v' = 13$ (see Figure 2).

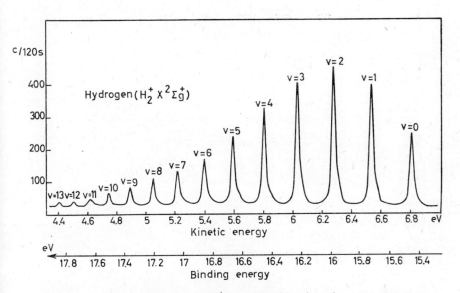

Figure 2 Vibrational structure in H_2^+ as observed by photoelectron spectroscopy with He 584 light (after Siegbahn *et al.*[34])

A measurement of the peak heights gives the relative transition probabilities, in very satisfactory agreement with the theoretically calculated Franck–Condon factors[2,59]. It is interesting to compare the He 584 spectrum with the photoelectron spectrum obtained with a variable wavelength as was done by Villarejo[55]. In this latter case, the equivalent of the positive ion photo-

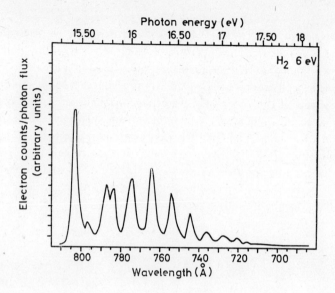

Figure 3   Photoionization curve of the hydrogen molecule (after Villarejo[55])

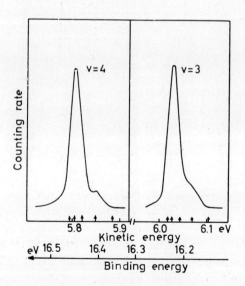

Figure 4   Rotational envelope of a vibrational band in H_2^+ as observed by photo-electron spectroscopy with He 584 light (after Siegbahn *et al.*[34])

ionization curve is obtained. Figure 3 reproduces Villarejo's curve, which is very different from Figure 2, in the same energy region, as a result of pre-ionization, particularly around the $v' = 0$ and $v' = 3$ transitions. Obviously, the relative intensities are completely different also. Figure 4 reproduces a detail of the $v' = 3$ and $v' = 4$ peaks with He 584 as obtained by Siegbahn *et al.*[34] to show that the asymmetry due to rotational structure is clearly detected. The vibrational spacings calculated from Figure 2 are in very good agreement with those calculated by the approximate anharmonic oscillator equation

$$E_{\text{vibr}} = hc\left[\omega_e\left(v + \tfrac{1}{2}\right) - \omega_e x_e\left(v + \tfrac{1}{2}\right)^2 - \tfrac{1}{2}\omega_e + \tfrac{1}{4}\omega_e x_e\right] \tag{5}$$

when using the spectroscopic data given by Herzberg[60].

The electron impact efficiency curve of hydrogen has also been the subject of many studies and much controversy. In principle the curve should be the result of direct ionization and preionization, so that no clear vibrational structure is to be expected. This was recently shown experimentally by McGowan *et al.*[62] and Lossing and Semeluk[63] who observed preionization humps in the efficiency curve, contrary to a previous study of Marmet and Kerwin[64].

3   Non-radiative and radiative transitions and the role of preionization

The density of excited neutral states accessible above the ionization limit is considerable and has been studied for a long time in UV spectroscopy[60,61]. In particular Rydberg states have been used for the calculation of IP's. In many cases, these states are preionized and, as such, may decay through an adiabatic transition towards the ionized state, most often vibrationally excited. These preionized states, as mentioned above, are very numerous and often observed in the photoionization curve as resonance peaks, in most cases narrow in energy, although very short lifetimes may result in uncertainty broadening.

In particular, the photoionization curves of diatomic molecules as O_2, N_2, CO, NO, show a great number of preionization peaks[26,28,65]. As a consequence, few really good vibrational frequencies have been obtained directly for the corresponding ions and use must be made of the photoelectron spectra, particularly with He 584[38,41,66]. These may be correlated, as far as intensities are concerned, with theoretical FC factors[41]. Wherever the comparison with spectroscopic values is possible, the agreement for vibrational spacings is excellent[66].

Anomalous vibrational intensity distributions are however observed when particular adequate wavelengths are used for irradiation. This is the case for the above mentioned molecules when Ne or Ar resonance light is used. Particularly striking is the case of oxygen which was studied by Collin and Natalis[66,67], Price[38] and Blake and Carver[68,69]. For instance we observed the O_2^+, $X^2\pi_g$ ground state up to $v' = 4$ with He 584 Å but up to $v' = 20$ with Ne 736–744 Å. In addition the spectra show a considerable intensity, variation being characterized, in the curve with Ne, by new broad maxima superimposed on the "normal" intensity distribution. The qualitative explanation is as follows: a preionized Rydberg state of O_2^* is situated at the energy of Ne 736 Å so that resonance absorption brings the O_2 into that Rydberg state which in turn preionizes to O_2^+, $X^2\pi_g$ producing a broad distribution of vibrationally excited O_2^+ ions. A quantitative treatment was recently

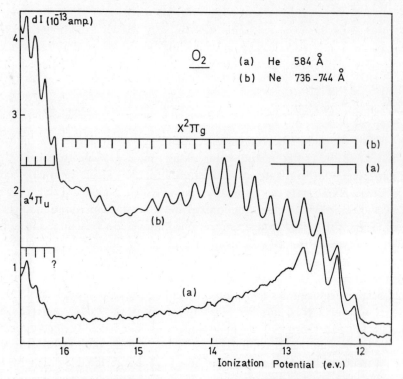

Figure 5   Photoelectron vibrational structure in the $X^2\pi_g$ ground state of O_2^+ as observed with different wavelengths: a) He 584; b) Ne 736–744, to show vibrational excitation enhancement due to preionization (after Natalis and Collin[66,67])

proposed by Blake and coworkers[69] who showed that the number of new broad maxima is equal to $n + 1$, where n is the vibrational quantum number of the Rydberg state concerned. The new features of the spectrum are then considered as due to the superposition of the indirect Franck–Condon preionization factors on the normal FC vibrational transition. It should be pointed out that the photoionization cross section may vary with wavelength so that an additional effect leading to an apparent relative increase of intensities for the high vibrational states is then observed.[39,66]

In polyatomic molecules, superexcitation and preionization are likely to occur, owing to the high density of available states. Such a general conclusion, however, does not seem confirmed experimentally, at least for hydrocarbons, as was recently shown by Chupka in the case of ethylene[71] (see also Dibeler *et al.*[70]). Chupka *et al.* suggest two possible reasons for this: predominance of very fast competing predissociations and slower preionization rates when the more loosely bound electrons are in delocalized orbitals, the "volume" of which reduces the probability of energy transfer between electrons. It is not clear however at the present time whether this is specific to hydrocarbons.

The situation is definitely different for small triatomic molecules such as CO_2, COS, CS_2, N_2O and H_2S. In these cases, the photoionization curves show a great number of preionization transitions in the energy interval between the first IP and about 21 eV*. In all those cases, Natalis and Collin[76,77,78] detected, in the photoelectron spectra obtained with various wavelengths, weak and diffuse bands not easily accounted for in terms of stable electronic ionic states. An analysis of the corresponding kinetic energies of the photoelectrons led to the conclusion that when the exciting radiation corresponds to a superexcited level, a transition to that level occurs. If the lifetime of this state and others in the vicinity happens to be long enough, a series of cascade internal non-radiative conversions may take place which eventually end up with a short-lived excited neutral state which preionizes into a vibrationally excited ion, the difference of energy between the final and original states being converted into the kinetic energy of the ejected electron.

Another process may give rise to apparently anomalous photoelectrons. Instead of neutral non-radiative transitions, radiative deactivation may occur from the initial superexcited state. In this case, the photoelectron ejection is accompanied by fluorescence. This picture was proposed by Blake and

---

* References: 72 for CO_2, COS, CS_2; 73 for N_2O; 74 for H_2S; for the photoelectron spectra: CO_2, COS, CS_2, N_2O: 75, 76, 77, 78; H_2S: 79.

Carver[68] to explain anomalies in the spectrum of oxygen as a function of wavelength. To our knowledge, this fluorescence has not yet been analysed, probably for lack of intensity.

IV PHOTOIONIZATION AND PHOTOELECTRON SPECTROSCOPY APPLIED TO FUNDAMENTAL AND STRUCTURAL PROBLEMS IN CHEMISTRY

1 Photoionization and the theory of mass spectra

The photoionization efficiency curves of molecules for the parent and fragment ions have been studied comprehensively by Inghram and collaborators particularly for hydrocarbons[80]. A thorough study of the structure of the curves, but in a small energy range, as well as of their derivatives has been made and a very complete and critical discussion of experimental results with respect to the theory of mass spectra (Quasi Equilibrium Theory—QET) was presented. Temperature effects, isotope effects, kinetic energy shifts, and metastable dissociations were investigated in this way and it was shown that the Quasi-Equilibrium Theory of mass spectra was to be modified, notably to account for a proper number of oscillators, in order to get agreement with experiment.

2 Molecular energy levels and potential energy curves

In favourable conditions, photoionization spectra may be very useful for the determination of ionization limits and vibrational spacings sometimes also of vibrational transition probabilities. From that standpoint, the photoelectron spectra are, however, more practical due to their simplicity. In the case of diatomic molecules, for instance, it may even be possible to reconstruct the potential energy curves from a consideration of the vibrational intensities and spacings. Many such attempts have been made recently; as an example, nitric oxide (NO) was carefully studied by Collin and Natalis[81] who detected no less than eight different ionic-electronic states below 21.21 eV and proposed a new potential energy diagram for NO^+. Photoelectron high resolution studies of NO by Price[38] and Lindholm[82] have also been published, including new and different energy diagrams. In the paper of Lindholm, the resolution was so good as to allow a very thorough study leading to an energy diagram considerably more complicated than anticipated by Collin and Natalis.

3   Nature of molecular orbitals and energy level correlations

Obviously, the removal of an electron in photoionization has very different effects on the resulting ion geometry according to whether the electron comes from a bonding, non-bonding or antibonding orbital. In particular, bonding and, to a lesser extent, anti-bonding electrons removed will cause the nuclear distances in the ion to be modified relative to that of the molecule, depending on the character of the Mo so that Franck–Condon factors are expected then to have appreciable values for high vibrational numbers. The corresponding photoelectron band, and to a certain extent the photoionization curve, will show a vibrational structure. On the contrary ionization on non-bonding electrons will show no extended vibrational structure. It should, therefore, be possible, by carefully examining the photoelectron curve, to obtain information on the nature of the molecular orbitals from which the electron was ejected. This has soon been recognized by Turner, Frost and Prince who applied those considerations to the discussion of the nature of molecular orbitals in many molecules. For instance, Frost *et al.*[83] and Price *et al.*[84] discussed in this way the IP's of hydrogen halides.

The systematic spectroscopic investigation of characteristic frequencies of molecules has since long been a very effective way of assigning experimental infra-red frequencies to particular vibrational motions in the molecule. It is worthwhile investigating whether, by analogy, the same kind of treatment is applicable to photoelectron spectra. Regularities are indeed found in homologous series in the position and shape of photoelectron bands, characteristic of a given type of molecular orbital. To get full use of such correlations, an extensive study of photoelectron spectra is necessary as in infra-red spectroscopy. This was done notably by Turner and collaborators for instance for benzene and its derivatives[85]. Many interesting features of the spectra may be correlated with molecular characteristics of the compounds; this includes notably the splitting of degenerate levels when removing degeneracy on substitution.

If a high resolution is available, it is to be expected that the fine structure of photoelectron spectra will be highly sensitive to the chemical nature of the molecule so that the spectrum may be considered, as in infra-red or ultra-violet, as a fingerprint of the compound and, therefore, makes photoelectron spectroscopy a simple but powerful tool for chemical analysis. In addition, the energy for ionization of a given type of electron may vary slightly according to substitution (nature and also position of the substituent) so that a real structural analysis may be based on such "chemical shifts". This observation was recently noticed by Turner[30] for the chloro- and bromobenzenes, but has been extensively used in a somewhat different way, to be described below, by Siegbahn and his collaborators[34].

The interpretation of experimental results often requires great care and use of many different techniques. A good example may be found in the photoionization and photoelectron spectra of benzene and its monosubstituted derivatives. The lowest ionization potential of benzene at 9.24 eV corresponds to the removal of an electron from the highest occupied orbital which is πe_{1g}. Substitution for instance in monofluorobenzene results in the splitting of this doubly degenerate level into 9.18 eV (πb_2) and 9.84 eV (πa_2) as observed by Momigny *et al.*[86], Al Joboury and Turner[87], Clark and Frost[88] and Natalis *et al.*[32]. It was assumed on the basis of previous helium photoelectron spectra[87,88] that no ionization potential of benzene was situated between 9.24 eV and the next spectroscopic Rydberg value at 11.48 eV, these two values being attributed to the π electrons. However, recent photoionization data[86] and photoelectron results obtained with Neon or Argon[32] have shown two more ionization potentials at 10.35 and 10.85 eV in benzene. These have been attributed by Momigny and Lorquet[89] to the σ orbitals σe_{2g} and σe_{1u} respectively, as partly evidenced by the degeneracy splitting of the 10.35 eV level into 9.41 and 10.35 eV in fluorobenzene. These conclusions are in agreement with recent theoretical calculations[90].*

4   Soft-X-ray photoelectron spectroscopy. The ESCA method†

X-Ray emission and absorption spectroscopy has been used for the study of the binding energies of atomic core electrons. However, recently Siegbahn and his collaborators[34] have developed a high resolution photoelectron method which allows the determination of binding energies to be done in the same way as discussed above by using soft X-ray irradiation of the sample. As such, the method has become at least as powerful as X-ray spectroscopy.

From an extended study of the soft X-ray photoelectron spectra of a great number of compounds, Siegbahn *et al.*[34] showed that these core binding energies are not only a property of the atom as such but also of its "chemical environment" or in other words of the molecular structure of the compound. In the same way, the charge distribution change in the valence shell accom-

---

* For a complete discussion, see 32, 86, 89, 91, 92.

† For detailed references, see 34 and 93.

panying a change of valence state produces also a modification of the binding energy and therefore should be detectable in a photoelectron experiment. Obviously these conclusions have the utmost importance since they potentially make photoelectron spectroscopy an analytical method as powerful as NMR. This is the reason for naming it the ESCA method (Electron Spectroscopy for Chemical Analysis).

Experience shows that the kinetic energies of electrons ejected from K-shells by soft X-rays are in the 1000 eV range and that chemical shifts of the

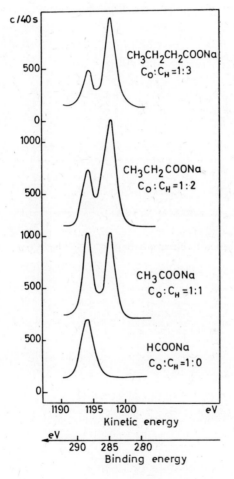

Figure 6   ESCA spectrum for sodium organic salts (after Siegbahn *et al.*[34])

order of a few volts are not uncommon, so that an energy resolution of the order of $1/10^3$ to $1/10^4$ should prove sufficient for detecting and measuring these shifts.

We shall show briefly how the application of the method may be extended to fields as varied as nuclear chemistry, inorganic chemistry, surface chemistry and biochemistry.

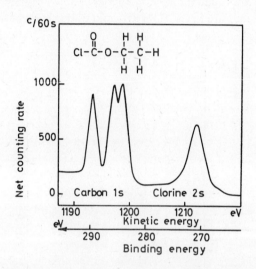

Figure 7   ESCA spectrum of ethylchloroformate showing the chemical shift in the carbon 1 s electron energy (after Siegbahn *et al.*[34])

Carbon atom 1 s electrons in a molecule have different binding energies according to their position and type of bonding. Inversely, the number of different types of carbon atoms may be deduced from the number of observed 1 s-carbon photoelectron bands or peaks. Figure 6 shows the cases of organic sodium salts where carbon lines for hydrocarbon and carboxyl groups appear distinctly in the 1200 eV range, whereas Figure 7 represents the ESCA spectrum of the ethyl chloroformate, where three types of carbon atoms are present and give rise to three lines in the 1190–1200 eV range, whereas chlorine 2 s electrons appear around 1213 eV.

Chemical shifts in the ESCA spectrum may be used to determine oxidation states of atoms. For instance 1 s-sulfur electrons show the shifts, by reference to sulfur itself, indicated in the following table[32].

Table 3 ESCA chemical shifts for 1s-sulfur electrons

Compound	Oxidation number	ESCA shifts (eV)	Compound	Oxidation number	ESCA shifts (eV)
sulfur	0	0	K_2SO_3	+4	+4.6
Na_2S	−2	−2.0	K_2SO_4	+6	+5.4
Na_2SO_3	+4	+4.5	Bz. SO_2Na	+2	+4.2
$Na_2S_2O_3$	+6	+5.3	Dextran.		
	−2	−1.7	OSO_3Na	+6	+5.6
Na_2SO_4	+6	+5.8			
K_2S	−2	−2.6			

Madelung constants and Hartree–Fock calculations based on the free-ion model may be performed and compared with ESCA results. In this way, the method is an invaluable means of obtaining new information for crystal field theory as applied to inorganic crystals. A very recent example is the evaluation of chemical effects on charge densities and core-electron binding energies in iodine and europium compounds by Fadley *et al.*[93].

In Figure 8 the ESCA spectra of dibenzyldisulfide and its corresponding thiosulfinate and thiosulfonate show again how easily the various types of sulfur atoms may be distinguished.

Considerable insight on the molecular and electronic structure of molecules may be gained. A good example is sodium azide (NaN_3). Since the value of the binding energy depends on the valence state, i.e. on the net charge of the "interested" atom. For instance, compounds containing nitrogen atoms with a single charge show a single nitrogen peak, whereas variously charged nitrogen atoms show a corresponding number of different peaks. Hexamethylenetetramine $(CH_2)_6N_4$ is a highly symmetrical molecule where all nitrogen atoms are equivalent and only one N peak is observed. On the contrary, NaN_3 is known to have variously charged nitrogen atoms, the central atom having a calculated positive charge of +0.64 whereas the other two N-atoms have a lower oxidation state with a net negative charge of −0.72, the general formula being $Na^+[N^- = N^+ = N^-]$. As a result, the ESCA spectrum (Figure 9) shows two well defined 1 s −N electron peaks around 1079 and 1084 eV.

Another interesting field of application of ESCA is the physical chemistry of surfaces. It is based on the observation that electrons in the 1000 eV range penetrate only thin layers of matter, of the order of a few hundred

11 Reed

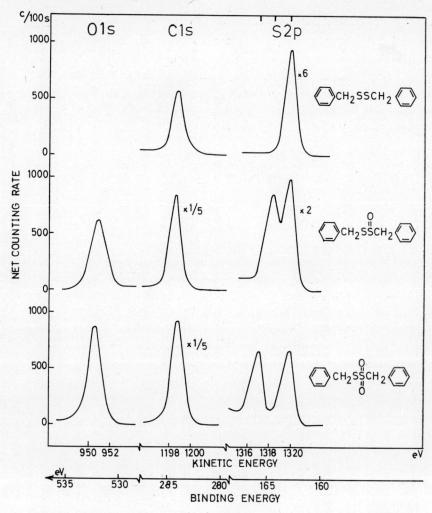

Figure 8   ESCA spectrum of dibenzyldisulfide, thiosulfinate and thiosulfonate showing the sulfur 2p electron chemical shift (after Siegbahn *et al.*[34])

angstroms. Therefore, it is expected that, according to the state of a surface, ESCA signals might be obtained for particular atoms which would disappear once the surface is covered by a layer of a given thickness. An experiment was performed by Siegbahn *et al.*[32] in which two multilayer samples containing a layer of α-bromostearic acid, which is 200 molecules thick, representing a thickness of about 8000 Å, were deposited on a chromium substrate. Under these conditions, no signal due to chromium was registered, proving complete shielding by the 8000 Å thickness, but ESCA lines were obtained for carbon 1 s-electrons and bromine 3 d-electrons of α-bromo-stearic acid. One of the samples was then covered by a double layer of stearic acid representing a thickness of about 40 Å. The carbon 1 s-line showed then no change, but the intensity of the bromine 3 d-electron line was very much decreased as well as slightly shifted in energy (Figure 10).

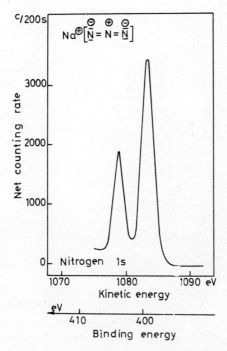

Figure 9   ESCA spectrum of NaN_3 showing the nitrogen 1s electron chemical shift (after Siegbahn *et al.*[34])

The method is thus useful for the study of molecular packing and of defects in monomolecular layers. It may be considered as able to yield information on surface problems concerning a thickness of *ca.* 100 to 500 Å.

Finally the ESCA spectroscopy, like the atomic emission spectroscopy, may be used as an analytical tool for the elemental analysis of samples and might prove interesting from that standpoint wherever the emission spectroscopy is not of easy application.

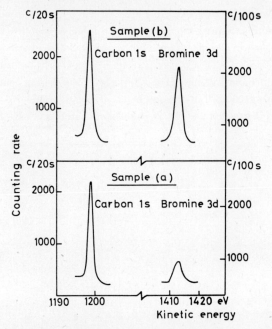

Figure 10   ESCA spectrum of 8000 Å multilayer samples of α-bromostearic acid adsorbed on a chromium surface: a) sample covered by a double layer (40 Å) of stearic acid; b) pure sample (after Siegbahn *et al.*[34])

CONCLUSIONS

To summarize, UV photoionization and UV photoelectron spectroscopy must be considered as important new experimental techniques for the study of molecular structure. In particular, they may help to clarify problems related to superexcitation, Rydberg levels, preionization and predissociation, Jahn–Teller effects, molecular orbital levels, electronic structure, and electronic states of ions, vibrational characteristics of ions. As for the ESCA method, we have shown how this new tool may be used for the investigation of a great variety of physicochemical subjects.

These techniques have reached such a degree of development that they now have acquired as great an importance for the physicist or the chemist as UV, IR or NMR spectroscopy, and represent a valuable additional technique for the investigation of fundamental problems.

References

1. J.E.Collin, in *Mass spectrometry*, edited by R.I.Reed, Acad. Press, London (1965), pp. 210–220.
2. M.E.Wacks, in *Modern Aspects of Mass Spectrometry*, edited by R.I.Reed, Plenum Press (1968), pp. 323–359.
3. J.A.R.Samson, in *Adv. Atom. and Molec. Physics*, edited by Bates and Esterman, vol. 2 177 (1966).
4. W.A.Chupka, J.Berkowitz, and K.M.A.Refaey, *J. Chem. Phys.* **50** 1938 (1969).
5. V.H.Dibeler and J.A.Walker, *J. Opt. Soc. Am.* **57**, 1007 (1967).
6. J.E.Collin, "Colloque sur l'Ionisation en Phase Gazeuse raréfiée, Université de Liège, 1963; *Bull. Soc. Chim. Belg.* **73**, 414 (1964).
7. E.Lindholn and coll., series of papers in *Ark. Fysik*; see for instance **39**, 65 (1969).
8. E.N.Lassettre and coll., *J. Chem. Phys.* **40** 1208 (1964); for more recent work see *J. Chem. Phys.* **48**, 5066 (1968).
9. J.E.Collin, in *Modern Aspects of Mass Spectrometry*, edited by R.I.Reed, Plenum Press, (1968) pp. 231–270.
10. C.E.Kuyatt, J.A.R.Simpson, and S.R.Mielczarek, *Phys. Rev.* **A385**, 138 (1965).
11. G.J.Schulz, *Phys. Rev.* **116**, 1141 (1959).
12. R.K.Curran, *J. Chem. Phys.* **38**, 780 (1963).
13. R.N.Compton, L.G.Christophotrou, G.S.Hurst and P.W.Reinhardt, *J. Chem. Phys.* **45**, 4634 (1966).
14. H.Ehrhardt and F.Linder, *Phys. Rev. Lett.* **21**, 419 (1968).
15. J.A.R.Samson, *Techniques of Vacuum Ultraviolet Spectroscopy*, Wiley, N.Y. (1967).
16. K.Codling and R.P.Madden, *J. Appl. Phys.* **36**, 380 (1965).
17. R.P.Madden and K.Codling, *Phys. Rev. Lett.* **10** 516 (1963); **12**, 106 (1964).
18. M.Seya, *Sci. Light* **2**, 8 (1952).
19. T.Namicka, *Sci. Light* **3**, 15 (1954).
20. K.Watanabe, *J. Chem. Phys.* **22**, 1564 (1954); **26**, 542 (1957).
21. F.P.Lossing and I.Tanaka, *J. Chem. Phys.* **25**, 1031 (1956).
22. H.Hurzeler, M.G.Inghram and J.D.Morrison, *J. Chem. Phys.* **28**, 76 (1958).
23. G.L.Weissler, J.A.R.Samson, M.Ogawa and G.R.Cook, *J. Opt. Soc. Am.* **29**, 339 (1959).
24. M.G.Inghram *et al.*, *J. Chem. Phys.* **34**, 179 (1961).
25. W.A.Chupka, J.Berkowitz *et col.*, *J. Chem. Phys.* **45**, 1287 (1966); **47**, 2921 (1967); **48**, 2337 (1968); **50**, 1938 (1969).
26. V.H.Dibeler and coll., *J. Res. Natl. Bur. Stds.* **68A**, 409 (1964); **71A**, 371 (1967); *J. Chem. Phys.* **43**, 1842 (1965); **44**, 1271 (1966); **45**, 1298 (1966); **48**, 4765 (1968); *J. Opt. Soc. Am.* **57**, 1007 (1967).
27. J.Momigny *et al.*, *Mem. Soc. Roy. Sci. Liège* XIII, 1 (1966); *J. Chim. phys.* **65**, 1213 (1968); *J. Mass Spectrom. Ion Phys.* **1**, 69 (1968).
28. G.R.Cook and B.K.Ching, *Aerospace Corpor. Techn. Report* TDR-469 (9260-01)-4, Los Angeles (1965).
29. G.V.Marr, *Photoionization Processes in Gases*, Ac. Press, N.Y. (1967).
30. D.W.Turner, in *Adv. Mass Spectrom.*, vol. IV, edited by E.Kendrick, Inst. Petrol. (1968).

31. D.W.Turner, in *Molecular Spectroscopy*, edited by P.Hepple, Inst. Petrol. (1968).
32. P.Natalis, J.E.Collin and J.Momigny, *J. Mass Spectrom. Ion Phys.* **1**, 414 (1968).
33. C.E.Brion and G.E.Thomas, *J. Mass Spectrom. Ion Phys.* **2**, 414 (1968).
34. K.Siegbahn, C.Nordling, A.Fahlman, R.Nordberg, K.Hanvein, J.Hedman, G.Johansson, T.Bergmark, S.E.Karlsson, I.Lindgren and B.Lindberg, *ESCA, Atominc, Molecular and Solid State Structure Studied by Means of Electron Spectroscopy*, Almqvist and Wiksells, Uppsala (1967).
35. F.I.Vilsesov, B.L.Kurbatov, and A.N.Terenin, *Doklad. Akad. Nauk SSSR* **138**, 1329 (1961).
36. D.W.Turner and M.I.Al Joboury, *J. Chem. Phys.* **37**, 3007 (1962).
37. H.J.Lempka, T.R.Passmore, and W.C.Price, *Proc. Roy. Soc.* **A304**, 53 (1968).
38. W.C.Price, in *Molecular Spectroscopy*, edited by P.Hepple, Inst. Petrol., London (1968).
39. D.C.Frost, C.A.McDowell, and D.A.Vroom, *Phys. Rev. Lett.* **15**, 612 (1965); *Proc. Roy. Soc.* **296A**, 566 (1967).
40. B.Brehm and A. von Puttkamer, *Z. f. Naturf.* **22a**, 8 (1967).
41. D.W.Turner and D.P.May, *J. Chem. Phys.* **45**, 471 (1966).
42. R.Spohr and E. von Puttkamer, *Z. f. Naturf.* **22a**, 705 (1967).
43. J.H.D.Eland and C.J.Danby, *J. Mass Spectrom. Ion Phys.* **1**, 111 (1968); *J. Sci. Inst. ser.2*, **1**, 406 (1968); *Z. f. Naturf.* **23a**, 355 (1968).
44. H.A.Bethe and E.F.Salpeter, *Quantum Mechanics of One- and Two-Electron systems*, Ac. Press, N.Y., 309 (1957).
45. J.Cooper and R.N.Zare, *J. Chem. Phys.* **48**, 942 (1968).
46. F.I.Vilesov, M.E.Akopyan, S.N.Lopatin, and V.I.Kheymenov, *Proc. Vth Int. Conf. Phys. Electronic and Atomic Collis.*, Leningrad, 606 (1967).
47. J.Berkowitz and H.Ehrhardt, *Phys. Lett.* **21**, 531 (1966).
48. J.Berkowitz, H.Ehrhardt, and T.Tekaat, *Z. f. Physik* **200**, 69 (1967).
49. D.C.Frost, personal communication, 1968.
50. J.A.R.Samson, *J. Opt. Soc. Am.* **59**, 356 (1969).
51. G.R.Cook and P.H.Metzger, *J. Opt. Soc. Am.* **54**, 968 (1964).
52. V.H.Dibeler, R.M.Reese, and M.Krauss, *J. Chem. Phys.* **42**, 2045 (1965).
53. W.A.Chupka and J.Berkowitz, *J. Chem. Phys.* **47**, 4320 (1967).
54. W.A.Chupka, M.E.Russell, and K.Refaey, *J. Chem. Phys.* **48**, 1518 (1968).
55. Don Villarejo, *J. Chem. Phys.* **48**, 4014 (1968).
56. H.Beutler and H.O.Junger, *Z. Physik* **100**, 80 (1936).
57. A.Monfils, *Mém. Roy. Belg. (Cl. Sci.)* **47**, 585 (1961).
58. M.I.Al Joboury and D.W.Turner, *J. Chem. Soc.*, 5141 (1963).
59. G.H.Dunn, *J. Chem. Phys.* **44**, 2592 (1966).
60. G.Herzberg, *Molecular Spectra and Molecular Structure. I. Diatomic Molecules*, Van Nostrand, 1950.
61. G.Herzberg, *Molecular Spectra and Molecular Structure. III. Electronic Spectra and Electronic Structure of Polyatomic Molecules*, Van Nostrand, 1966.
62. J.W.McGowan, M.A.Fineman, E.M.Clarke, and H.P.Hanson, *Phys. Rev.* **167**, 52 (1968).
63. F.P.Lossing and G.P.Semeluk, *J. Mass Spectrom. Ion Phys.* **2**, 408 (1969).
64. P.Marmet and L.Kerwin, *Can. J. Phys.* **38**, 972 (1960).

65. A.J.C.Nicholson, *J. Chem. Phys.* **39**, 93 (1964).
66. J.E.Collin and P.Natalis, *J. Mass Spectrom. Ion Phys.* **2**, 231 (1969).
67. P.Natalis and J.E.Collin, *Chem. Phys. Lett.* **2** 414 (1968).
68. A.J.Balake and J.H.Carver, *J. Chem. Phys.* **47**, 1038 (1967).
69. J.L.Baker, A.J.Blake, J.H.Carver, and V.Kumar, *J. Quant. Spectr. and Radiat. Transfer* (submitted), personal communication, 1968.
70. R.Botter, V.H.Dibeler, J.A.Walker, and H.M.Rosenstock, *J. Chem. Phys.* **45**, 1298 (1966).
71. W.A.Chupka, J.Berkowitz, and K.M.A.Refaey, *J. Chem. Phys.* **50**, 1938 (1969).
72. V.H.Dibeler and J.A.Walker, *J. Opt. Soc. Am.* **57**, 1007 (1967).
73. V.H.Dibeler, J.A.Walker, and S.K.Liston, *J. Res. Natl. Bur. Stds.* **71A**, 371 (1967).
74. V.H.Dibeler and S.K.Liston, *J. Chem. Phys.* **49**, 482 (1968).
75. C.R.Brundle and D.W.Turner, *J. Mass Spectrom. Ion Phys.* **2**, 195 (1969)
76. J.E.Collin and P.Natalis, *J. Mass Spectrom. Ion Phys.* **1**, 121 (1968).
77. P.Natalis and J.E.Collin, *J. Mass Spectrom. Ion Phys.* **2**, 221 (1969).
78. J.H.D.Eland and C.J.Danby, *J. Mass Spectrom. Ion Phys.* **1**, 111 (1968)
79. P.Natalis and J.E.Collin, *20ème Réunion annuelle de la Société de Chimie physique*, Paris, mai 1969; *J. Chim. phys.* (to appear).
80. B.Steiner, C.F.Giese, and M.G.Inghram, *J. Chem. Phys.* **34**, 189 (1961).
81. J.E.Collin and P.Natalis, *J. Mass Spectrom. Ion Phys.* **1**, 483 (1968).
82. E.Lindholm, *Ark. f. Fysik*, in press (personal communication, 1969).
83. D.C.Frost, C.A.McDowell, and D.A.Vroom, *J. Chem. Phys.* **46**, 4255 (1967).
84. H.S.Lempka, T.R.Passmore, and W.C.Price, *Proc. Roy. Soc.* **A304**, 53 (1968).
85. A.D.Baker, D.P.May, and D.W.Turner, *J. Chem. Soc.* **B**, 22 (1968).
86. J.Momigny, C.Goffart, and L.D'Or, *J. Mass Spectrom. Ion Phys.* **1**, 53 (1968).
87. M.I.Al Joboury and D.W.Turner, *J. Chem. Soc.*, 4434 (1964).
88. I.D.Clark and D.C.Frost, *J. Am. Chem. Soc.* **89**, 244 (1967).
89. J.Momigny and J.C.Lorquet, *Chem. Phys. Lett.* **1**, 505 (1968).
90. J.M.Schulman and J.W.Moskowitz, *J. Chem. Phys.*, **47**, 3492 (1967).
91. A.D.Baker, C.R.Brundle, and D.W.Turner, *J. Mass Spectrom. Ion Phys.* **1**, 443 (1968).
92. J.Momigny and J.C.Lorquet, *J. Mass Spectrom. Ion Phys.* **2** (1969), in press.
93. C.S.Fadley, S.B.M.Hagstrom, M.P.Klein, and D.A.Shirley, *J. Chem. Phys.* **48**, 3779 (1968).

The study of rearrangement reactions by field ionization mass spectrometry

H. D. BECKEY and K. LEVSEN

Institut für Physikalische Chemie der Universität Bonn, Germany

---

I THEORETICAL INTRODUCTION

Rearrangement reactions of organic ions have been studied extensively with electron impact mass spectrometry by a number of authors during the past few years. Williams and Cooks[1] point out that lowering the energy of the ionizing electrons from 70 eV to a value of slightly above the appearance potential of the rearranged ions increases in a number of cases the intensity of these ions relative to other fragment ions formed by direct bond rupture from the same molecular ions. This phenomenon is observed because both the activation energy and the frequency factors are small with many rearrangement reactions.

Following the arguments of Williams *et al.*[2], the situation is illustrated in Figure 1a. The rate constant k for the unimolecular decomposition of the molecular ion is plotted as a function of the internal energy of the ions. The shapes of the curves are taken arbitrarily, and no reference is made to specific formulations of the quasi-equilibrium theory (QET) of mass spectra[3]. The function $k_{(E)}$ has been determined experimentally with a special EI (Electron Impact) method by Hertel, Osberghaus and Ottinger[4-6] for a number of substances. A lower limiting value is imposed on the measurement of reaction rates by the geometry and the potential distribution of conventional electron impact (EI) ion sources. The draw-out time for ions from normal EI sources is about 10^{-6} sec. Therefore, fragment ions recorded at the nor-

169

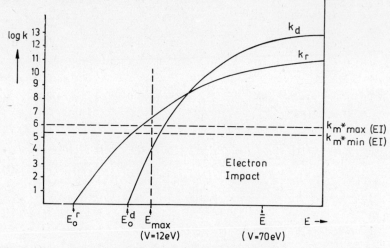

Figure 1a $\log k$ versus internal energy E of the molecular ions (schematically). Horizontal dashed lines: Boundary conditions for EI $\log k = 5.5 - 6$: Metastable decomposition. $\log k = 6 - 13$: normal fragmentation. E_0^r and E_0^d = activation energies for rearrangement and direct bond fission, respectively. E_{max} = maximum energy transferred by an electron with kinetic energy $V = 12$ eV. $\overline{E}$ = average energy transferred by an electron with kinetic energy $V = 70$ eV. k_m = rate constant for metastable decomposition

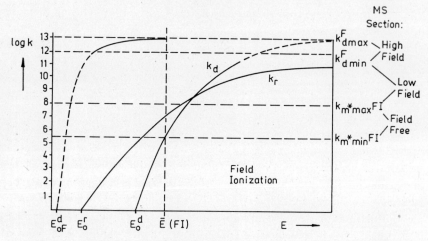

Figure 1b $\log k$ versus internal energy E of the molecular ions (schematically). Horizontal dashed lines: Boundary conditions for FI. $\log k = 12 - 13$: Field induced direct bond fission with reduced activation energy E_{0F}^0 $\log k = 5.5 - 8$: Metastable decomposition in the field free space. $\overline{E}$ (FI) = average energy transferred to the molecular ion by field ionization. E_0^r and E_0^d = activation energies for rearrangement and direct bond fission, respectively. The high, low and zero field sections of the mass spectrometer are indicated in the figure

mal mass positions relative to the molecular ions are produced within the time interval between one molecular vibrational period (about 5×10^{-14} sec) and 10^{-6} sec, the slower processes being more probable than the faster ones.

Inspecting Figure 1a one has to realize that only processes attributed to k-values greater than 10^6 can contribute to a normal fragment peak. (Horizontal line at $\log k = 6$ in Figure 1a). The average internal energy E transferred to a molecular ion is of the order of several eV if the energy V of the impingeing electrons is 70 eV. The average internal energy is reduced, in general, to a few tenths of an eV if the energy of the electrons is reduced to a value of slightly above the appearance potential of the fragment ions (12 eV, for example).

The impact of this reduction of internal energy E on the ratio of the intensities I_r/I_d (I_r = intensity of the fragment formed by a rearrangement process, I_d = intensity of a fragment formed by a direct bond fission) may be discussed by means of Eqn. (1).

$$k = \nu \left(\frac{E - E_0}{E} \right)^{s-1} \tag{1}$$

ν = frequency factor, E = internal energy of the parent ion, E_0 = activation energy, s = number of vibrational modes.

Equation (1) describes the rate constant by the simplest version of the QET[3] for the case of a molecule with no internal rotations. In a great number of rearrangement reactions internal rotations of the molecular ions are stopped in the activated complex. Rosenstock *et al.*[3] have derived a rate equation for this case, which, on the same first level of approximation as Eq. (1), is given by

$$k = \nu' \left(\frac{E - E_0}{E} \right)^{s-(L/2)-1} (E - E_0)^{(L-L^*)/2} \tag{2}$$

where ν' is a frequency factor depending on the vibrational and rotational molecular frequencies, E, E_0 and s have the same meaning as before and L and L^* are the numbers of rotational modes of the parent ion and the activated complex, respectively. Equation (2) describes the slower rise of k with E in a qualitatively right manner for the low energy region. However, the equation leads to infinitely large k-values if the energy is steadily increased. This is physically unreasonable, and therefore, Eqn. (2) has to be replaced by another power series of $(E - E_0)/E$. Because of lack of an exact equation, let us put for qualitative discussions

$$k = \nu_r f_{r((E-E_0)/E)} \tag{3}$$

where f_r is a power series of the argument $(E - E_0)/E$, depending on the number of oscillators and the number of stopped internal rotations for a rearrangement reaction. The function f_r is such that k increases less steeply with the energy for the case of a rearrangement reaction than for the case of direct bond fission (see Figure 1). The constant v_r of Eqn. (3) also reflects the effect of stopped rotations which may reduce v_r by several orders of magnitude as compared to v of Eqn. (1). In a general sense we may speak of a "low frequency factor" for a rearrangement reaction, keeping in mind that the reduction of v is only to some extent due to the reduction of actual molecular frequencies, but mainly due to the stopping of rotations.

Regardless of the exact formulation of this equation one can qualitatively derive from Eqn. (3) the effect of lowering the value of E.

Let us assume in accordance with Williams and Cooks[1] that both E_0 and v are much smaller with a rearrangement than with a direct bond fission process. Then from the condition $E \gg E_0$ the function $f_{r(E-E_0)/E}$ is near unity, and the ratio of the rate constants $k_r/k_d \simeq v_r/v_d$ for the rearrangement and the direct bond fission processes. The intensity ratio of two competing decomposition processes for a given small energy interval is equal to the ratio of the rate constants. Therefore, I_r/I_d should be small at $E \gg E_0$. If E is reduced to a value slightly above E_0, the factor $(E - E_0)/E$ will be rate determining, and $k_r/k_d > 1$, as can be seen from the figure, and hence $I_r/I_d > 1$. Table 1 represents the $(I_r/I_d \,(12 \text{ eV}))/(I_r/I_d \,(70 \text{ eV}))$ values for some rearrangement reactions studied by the present authors.

Let us now turn to the case of field ionization. It is well known that only a small amount of energy is transferred to molecules in the case of field ionization (F.I.). Therefore, one should expect conditions similar to low energy electron impact. The $k_{(E)}$ versus E curves, according to the QET, should be independent of the method of ionization, i.e. they should be the same for *EI* and *FI* in a time interval where the field is too low to change the potential curves appreciably.

On the other hand, the authors have pointed out earlier[7] that the time scales for observation of normal fragment peaks are entirely different in the cases of field ionization and electron impact ionization. Dissociation must occur within a time interval of about $5 \times 10^{-14} - 10^{-12}$ sec in the case of field ionization if the fragment ions are to appear at the normal mass position as compared to the molecular ions.

Molecular ions decomposing later than 10^{-12} sec after field ionization have already arrived in a region where the electric potential is appreciably lower

Table 1

Compound	Processes		m/e	$\dfrac{(I_r/I_d) \text{ low eV*}}{(I_r/I_d) \text{ 70 eV}}$ E.I
Acetic acid phenylethyl ester	$M^+ - CH_3 \cdot CO \cdot OH$	(r)	(104)	2.6
$CH_3 \cdot CO \cdot OCH_2CH_2C_6H_5$	$M^+ - CH_3 \cdot CO \cdot O$	(d)	(105)	
Benzoic acid butyl ester	$(C_6H_5 \cdot CO \cdot OH_2)^+$	(rr)	(123)	2.3
$C_6H_5 \cdot CO \cdot OC_4H_9$	$(C_6H_5 \cdot CO \cdot OH)^+$	(r)	(122)	
Butyric acid-n-amyl ester	$M^+ - C_3H_7 \cdot CO \cdot OH$	(r)	(70)	4.0
$C_3H_7 \cdot CO \cdot OC_5H_{11}$	$M^+ - C_3H_7 \cdot CO \cdot O$	(d)	(71)	
	$(C_3H_7 \cdot CO \cdot OH_2)^+$	(rr)	(89)	1.3
	$(C_3H_7 \cdot CO \cdot O)^+$	(d)	(87)	
	$(C_3H_7 \cdot CO \cdot OH)^+$	(r)	(88)	1.5
	$(C_3H_7 \cdot CO \cdot O)^+$	(d)	(87)	
	$M^+ - C_2H_4$	(r)	(130)	2.1
	$M^+ - C_2H_5$	(d)	(129)	
Valerophenone	$M^+ - C_3H_6$	(r)	(120)	4.2
$C_6H_5 \cdot CO \cdot C_4H_9$	$M^+ - C_3H_7$	(d)	(119)	
Salicylic acid methyl ester	$M^+ - CH_3OH$	(r)	(120)	2.8
$C_6H_5(OH) \cdot CO \cdot OCH_3$	$M^+ - CH_3O$	(d)	(121)	
2-Hydroxybenzyl alcohol	$M^+ - H_2O$	(r)	(106)	12.5
$C_6H_5(OH) \cdot CH_2OH$	$M^+ - OH$	(d)	(107)	
p-Nitrobenzene sulfonamide	$M^+ - SO_2$	(r)	(138)	17.5
$NO_2 \cdot C_6H_5 \cdot SO_2 \cdot NH_2$	$M^+ - NH_2$	(d)	(186)	
Diphenylether	$M^+ - CO$	(r)	(142)	9.0
$C_6H_5 \cdot O \cdot C_6H_5$	$M^+ - C_6H_5O$	(d)	(77)	
Dimethyldisulfide	$M^+ - HS$	(r)	(61)	6.8
CH_3SSCH_3 (94)	$M^+ - CH_3$	(d)	(79)	
Diphenylsulfoxide	$M^+ - SO$	(r)	(154)	11.4
$C_6H_5 \cdot SO \cdot C_6H_5$	$M^+ - C_6H_5$	(d)	(125)	

* 11–12 eV

r = rearrangement; rr = double rearrangement; d = direct bond rupture

than at the field producing tip (see Figure 2). Therefore, the energy eU^* of the fragment ions entering the mass spectrometer is lower than that of the fragments formed within 10^{-12} sec after FI:

$$eU^* = e\,(U_0 - U)\,m/M + eU, \qquad (4)$$

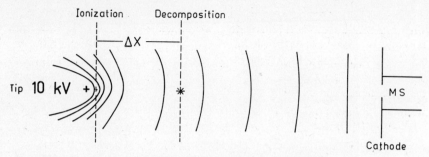

Figure 2   Schematic representation of equipotential lines between a tip anode and a cathode

where U_0 = tip potential, U = potential at the place of decomposition, m = fragment ion mass, M = molecular ion mass; e = unit charge. The kinetic energy of the parent ions is eU_0. The delayed fragment ions appear at an apparent mass number $m^* < m$ in the mass spectrum where

$$m^* U_0 = mU^* \tag{5}$$

substituting (4) into (5), one obtains:

$$m^* = m \left[\frac{m}{M} + \frac{U}{U_0} \left(1 - \frac{m}{M} \right) \right] \tag{6}$$

Now inspecting Figure 1b, one can realize that the $k_{(E)}$-curve corresponding to a process with a low frequency factor never reaches the boundary condition $k_{(E)} > 10^{12}$ which is a necessary condition for observation of a fragment at its normal mass position. In the case of FI. $k_{(E)} > 10^{12}$ is not arrived even at much larger energies than transferred to the molecule in reality. The limit $k = 10^{12}$ depends to some extent on the resolving power of the mass spectrometer[8]. These considerations lead to the general result: *Reactions with a small frequency factor ($k <$ several $10^{11}\ sec^{-1}$) occurring in the gas phase cannot be detected in an FI mass spectrum by fragments which have peak maxima at "normal" mass positions.*

"Normal" mass position means an m/e value which exactly fits with a mass scale calibrated with two parent ion masses near that value. These slow reactions can be detected, however, either as "fast metastable" fragment ions which are formed between the FI emitter and the cathode with shifted peak maxima at m_f^* ($m > m_f^* > m^2/M$), or as "normal metastable" ions ($m_n^* = m^2/M$).

In contrast to these conditions for field ionization, very intense fragment ions originating from reactions with low frequency factors can be observed at the normal mass position in the electron impact mass spectra. By lowering the electron energy—as stated before—the abundance of these fragment ions will often be larger than that of fragments originating from direct bond fission processes. For these reasons it is interesting to compare for the case of *FI* and *EI* the ratio of the intensities of fragment ions originating from low frequency factor reactions, normalized to the fragment sum intensity, $I_{\Sigma f}$. A ratio R can be studied which is defined by

$$R = \frac{(I_{nl}/I_{\Sigma f})_{FI}}{(I_{nl}/I_{\Sigma f})_{EI}} \tag{7}$$

where I_n are the intensities of the peak *maxima* at "normal" mass positions, and the index l refers to a low frequency factor. A rearrangement reaction with a small maximum rate constant (say $k = 10^{11} \ \text{sec}^{-1}$) leading to a shifted mass peak may have a non-vanishing peak tail at the normal mass position. This is because of the unimolecular decomposition kinetics: a small fraction of molecules associated with $k = 10^{11} \ \text{sec}^{-1}$ decompose already in the time interval $10^{-14} - 10^{-12}$ sec. The intensity of this peak tail at the normal mass position is not of interest here, because we try to evaluate the maximum rate constants from the shifted peak *maxima*. Therefore, in evaluating the mass spectra for calculation of R, only the difference between a peak maximum at normal mass position and the tail intensity of a shifted peak at the same position has to be taken, as shown in Figure 3a.

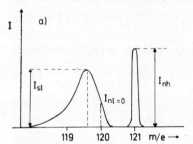 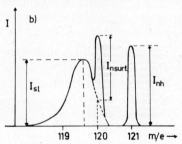

Figure 3a Fragment peak with a shifted peak maximum at $m/e = 119.6$ originating from a reaction with a low frequency factor $v_r < 10^{11} \ \text{sec}^{-1}$. Intensity of the shifted peak maximum: I_{sl}. There is no peak maximum at the normal mass position, i.e. $I_{nl} = 0$, $R = 0$. The peak at $m/e = 121$ is due to a direct bond fission (high frequency factor)

Figure 3b Fragment peak of the same gas phase reaction, superimposed by a surface rearrangement peak of intensity $I_{n\,surf}$. R has a large value for this special case

The ratio R should be zero, or at least very small according to these considerations. It should be kept in mind, however, that these statements are applicable only to gas phase reactions. Reactions with low frequency factors can occur on the surface of the field ion emitter if the average adsorption time is large enough. These reactions do not fall into the range of application of the above mentioned rule, and thus the R-factors may become large in these cases (see Figure 3b).

A class of reactions which have small frequency factors in many cases[2] are the *rearrangement reactions* of organic ions in the gas phase. It should be remembered however, that rearrangement reactions comprise a large range of frequency factors depending on the nature of these processes. If the migration of a proton, for example, is favoured by a sterically preferred configuration of the molecular ion, the frequency factor will be slightly reduced below the average value for direct bond fission processes. If, on the other hand, the rearrangement process is a complicated one, i.e. if the entropy of activation of the corresponding thermal rearrangement reaction has a large negative value, the frequency factor can be reduced by several orders of magnitude.

Another interesting feature concerning the normal metastable ions formed by *FI* may be derived from Figure 1b. The figure shows the minimum and the maximum time of flight of the ions entering and leaving the field free space between the mass spectrometer entrance slit and the magnetic field. (Figure 1b is based on the conditions for an ion source without a lens system, i.e. without retardation of the ions.) The residence time of the metastable ions in this field free space extends from about 10^{-8} to 3×10^{-6} sec, whereas this time interval is only about 1×10^{-6} to 3×10^{-6} sec in the case of electron impact. Both because of this reason and because the fraction of low energy ions is larger with *FI* than with *EI*, the relative metastable ion intensity is larger in the case of *FI* than in the case of *EI*.

It can be derived further from Figure 1b that the relative normal metastable ion intensities should be the larger the lower the activation energy and the frequency factor are. Therefore, a factor Q is introduced which is equal to the ratio of the intensities of the integrated fast metastable peak (I_{if}) and the corresponding normal metastable peak (I_{im*}):

$$Q = I_{if}/I_{im*} \tag{8}$$

The Q-factor should be small for slow gas phase reactions.

For a comparison of *FI* and *EI* mass spectra with respect to kinetic problems one should study rearrangement reactions which yield very intense fragment peaks in the low voltage *EI* spectra. The intensity of many rearrangement peaks increases with decreasing electron energy relative to the other fragments because of the low activation energy. The elimination of small molecules of low energy content such as CO, CO_2, SO, SO_2, C_2H_4, H_2O etc. is the driving force for such reactions.

Some rearrangement processes are energetically unfavourable, especially if they are double step processes like the rearrangement of *p*-anisidine.

$$m/e = 123 \qquad m/e = 108 \qquad m/e = 80$$

The 12 eV *EI* mass spectrum shows a peak at $m/e = 80$ of only 0.1% rel. abundance. (27% at 70 eV). This fragment is not found in the *FI* mass spectrum. Therefore, the studies have been directed mainly to single step rearrangement processes, although intense multistep rearrangement processes are not unusual in the *EI* mass spectra.

II RESULTS

The reactions which have been studied are grouped in the following way:

a) Rearrangement reactions with hydrogen transfer

b) Skeletal rearrangement reactions.

The *FI* mass spectra are drawn in logarithmic intensity scale because of the low fragment intensities in contrast to the *EI* mass spectra which are plotted with a linear intensity scale. The most important metastable peaks are plotted in the *FI* mass spectra by dashed lines placed at the measured mass positions. The processes leading to these peaks are also indicated.

R-values are formed according to Eqn. (7), i.e. the intensities are related to the peak *maxima* at "normal" mass position. The *R*-factors of most of the rearrangement reactions are extremely small, or zero, except for a few cases of surface rearrangement reactions. Further a factor *S* is formed which refers to the integrated peak intensity of the—normally shifted—rearrangement peak, I_{ir}:

$$S = \frac{(I_{ir}/I_{\Sigma f})_{FI}}{(I_{ir}/I_{\Sigma f})_{EI}} \tag{9}$$

($I_{\Sigma f}$ is the sum of the fragment ion intensities).

There are no principal theoretical restrictions for the integrated intensities of the shifted rearrangement peaks, and therefore, S may become greater than one, in contrast to the R-factors which are much smaller than one for gas phase reactions.

These S-factors are formed in order to show a certain parallelism to the Q-factors (defined by Eqn. (8)). A reaction having a very small maximum rate constant will show an intense normal metastable peak and a weak fast metastable peak (ion residence time about $10^{-12} - 10^{-9}$ sec). Hence both S and Q should be small for that case. By the same argument a fast rearrangement reaction should lead to both large S and Q factors. These factors are listed in Tables 2–3, and it is seen that there is a certain parallelism with respect to the order of magnitude of S and Q. The S and Q factors are classified as "low" *(l)* if they are smaller than 0.1 and as "high" *(h)* if they are larger than 0.1. Most of the reactions can be recognized as either *ll* or *hh*. There are also a few intermediate cases *lh* (salicylic acid methylester and 2-hydroxybenzylalcohol) where the rates are between those of very fast and very slow rearrangement reactions.

II.1   Rearrangement reactions with hydrogen transfer

One of the most extensively studied reactions of this type in the EI mass spectrometry is the McLafferty rearrangement. The maximum rate constant for the transfer of a γ-hydrogen atom to the carbonyl oxygen atom of a ketone, for example, can be derived qualitatively in the following way. The normal frequency factor of a C–C single bond vibration in an aliphatic compound is about 3×10^{13} sec^{-1}. In a McLafferty rearrangement the migrating H-atom is brought into the vicinity of the heteroatom by torsions of parts of the molecules around the C–C axes. This motion is a hindered rotation with a frequency factor of about 6×10^{12} sec^{-1}, which corresponds to a period of 1.6×10^{-13} sec. It can be derived from stereochemical models that a H-atom is brought into a sterically favourable rearrangement position only after several torsional periods, on the average. The small activation energy of hydrogen rearrangement reactions further reduces the rate constants. Estimating that both the energetic and the steric factors together reduce the rate constants by at least a factor of ten we arrive at an average rearrangement time of $\tau_R > 10^{-11}$ sec.

According to these considerations, one should not observe McLafferty rearrangement peaks in the *FI* mass spectra with peak maxima at the "normal" mass positions if the reactions occurred in the gas phase, and consequently the *R* factors should be small.

The experimental results show that the *R*-factors are small in fact with most of the McLafferty or related rearrangements, but that they are much larger than expected in a few cases. It will be shown that the reactions are not pure gas phase reactions in the latter case. Now a number of rearrangement reactions with hydrogen transfer will be discussed in some detail.

II.1.1 *McLafferty rearrangements of ketones*

These rearrangements have been studied for the series butyrophenone, valerophenone and caprophenone.

The main reaction is a single hydrogen rearrangement leading to a fragment $m/e = 120$. A rearrangement peak $m/e = 120$ from butyrophenone of a few percent abundance has been reported previously by Chait *et al.*[16]

Figure 4 shows the *EI* and *FI* mass spectra of valerophenone. One of the most interesting features of the *FI* mass spectrum is the shift of the maximum of the rearrangement peak $m/e = 120$ by several tenths of a mass unit towards smaller mass numbers. The peak shift depends on the geometry and the radius of curvature of the emitter. A wire emitter with a radius of curvature of ~ 0.0025 mm has always been used. Moreover the peak shift depends on the ratio of the emitter voltage ($+ U_{an}$) to the total voltage ($+ U_{an} - U_{cath}$). Therefore, the peak shifts listed afterwards are related to certain $+ U_{an}$ and $- U_{cath}$ values, which are 3 and 7 kV, respectively, for the present case.

	Δm in mass units	
butyrophenone	-0.47	± 0.08
valerophenone	-0.55	± 0.08
caprophenone	-0.71	± 0.08

Table 2   Rearrangements with H-transfer

Compound (Molecular weight)	Process	I (%) F.I.	I (%) E.I.(12 eV)	$I/\Sigma I_{fragm}$ (%) F.I.	$I/\Sigma I_{fragm}$ (%) E.I.(12 eV)	I (%) F.I. Metast.	s	Q
Butyrophenone (184)	$M^+ - C_2H_4$ (120) (McLafferty R.)	0.33 (150°)	49 (250°)	8	20	1.2	0.40	0.27
Valerophenone (162)	$M^+ - C_3H_6$ (120) (McLafferty R.)	2.9 (150°)	100 (150°)	26	50	6.3	0.52	0.46
Caprophenone (176)	$M^+ - C_4H_8$ (120) (McLafferty R.)	3.1 (150°)	100	33	76	4.6	0.44	0.67
Menthone (154)	$M^+ - C_3H_6$ (112) (McLafferty R.)	0.03 (25°)	100 (150°)	<1.4	48	0.45	0.03	0.07
Benzoic acid butyl ester (178)	C_6H_5—$COOH_2^+$ (123) "two hydrogen transfer"	<0.05 (25°)	100 (120°)	2.5	71	8.1	0.035	<0.006
	$M^+ - OH$ (161)	<0.01 (90°)	0.9 (120°)	<0.15	0.6	0.3	<0.25	<0.03
Benzoic acid propyl ester (164)	C_6H_5—$COOH_2^+$ (123) "two hydrogen transfer"	0.05 (25°)	100 (120°)	6	70	7.2	0.08	0.007
Butanoic acid-n-amyl ester (158)	C_3H_7—$CH=CH^+$ (70) (McLafferty R.)	38.0 (130°)	100 (150°)	20	36	0.4	0.55	95
	C_3H_7—$COOH_2^+$ (89)	2.5 (130°)	74 (150°)	1.2	26	0.6	0.05	4
	$M^+ - C_2H_4$ (130) (McLafferty R.)	0.2 (130°)	2.7 (150°)	<0.1	1.0	0.5	0.1	0.4
Butanoic acid butyl ester (144)	C_2H_9—$CH=CH_2$ (56) (McLafferty R.)	100 (90°)	77 (100°)	87	29	0.8	3.0	125

Compound	Ion							
Butanoic acid butyl ester (144)	C₃H₇—COOH₂⁺ (89) "two hydrogen transfer"	0.8 (90°)	100 (100°)	0.6	38	15	0.016	0.05
	M⁺ − C₂H₄ (116) (McLafferty R.)	0.1 (90°)	4.7 (100°)	<0.08	1.8	6	0.04	0.02
Butanoic acid propyl ester (130)	CH₃—CH=CH₂⁺ (42) (McLafferty R.)	100 (100°)	8.0 (110°)	81	3.5	—	23	—
	C₃H₇—COOH₂⁺ (89) "two hydrogen transfer"	1.3 (100°)	100 (110°)	1.0	45	7.4	0.022	0.18
	M⁺ − C₂H₄ (102) (McLafferty R.)	0.1 (100°)	8.5 (110°)	<0.08	3.9	4.8	0.02	0.02
Acetic acid phenylethyl ester (164)	C₆H₅—CH=CH⁺ (104) (McLafferty R.)	100 (150°)	100 (150°)	67	90	11	0.75	9
Salicylic acid methyl ester (152)	M⁺ − CH₃OH (120) (Orthoeffect)	0.03 (50°)	46 (140°)	4	90	0.13	0.04	0.2
2-Hydroxybenzyl alcohol (124)	M⁺ − H₂O (106) (Orthoeffect)	0.06 (100°)	65 (110°)	4	68	0.25	0.06	0.2
p-Cresol (108)	M⁺ − H₂O (90)	<0.01 (150°)	1.0 (160°)	<1.2	9	0.12	<0.13	<0.08
Diphenylsulfoxide (202)	M⁺ − H₂O (184)	<0.01	0.6 (130°)	<0.2	0.8	0.11	<0.25	<0.09

Table 3 Skeletal rearrangements

Compound (Molecular weight)	Process	I (%) F.I.	I (%) E.I. (12 eV)	$I/\Sigma I_{Fragm}$ (%) F.I.	$I/\Sigma I_{Fragm}$ (%) E.I. (12 eV)	I (%) F.I. Metast.	s	Q
Diphenylsulfoxide (202)	$M^+ - CO$ (174)	<0.01	4.8 (135°)	<0.2	8	0.11	<0.02	<0.1
	$M^+ - SO$ (154)	<0.01	16 (135°)	<0.2	27	0.14	<0.007	<0.07
p-Cresol (108)	$M^+ - CO$ (80)	<0.01 (150°)	1.6 (160°)	<1.3	15	0.3	<0.09	<0.03
Diphenylether (170)	$M^+ - CO$ (142)	<0.001 (~350°)	2.3* (150°)	<0.2	55*	0.5	<0.004	<0.01
Dimethyldisulfide (94)	$M^+ - HS$ (61)	0.005 (~350°)	8 (140°)	1	41	0.08	0.02	0.06
Dimethylacetylene dicarboxylate (142)	$M^+ - CO_2$ (98)	<0.005 (25°)	51 (150°)	<0.5	11	1.8	<0.05	0.003
Anthraquinone (208)	$M^+ - CO$ (180)	<0.01 (290°)	8 (220°)	<0.5	68	0.35	<0.007	<0.03
N,N-Dimethyl-N'-phenyl-formamidine (148)	$M^+ - H_2CN$ (120)	<0.001 (25°)	8* (150°)	<0.1	8*	0.04	<0.01	0.03
	$M^+ - CH_3$—HCN (106)	0.001 (25°)	9* (150°)	0.1	10*	–	0.01	–
p-Nitrobenzenesulfon-amide (202)	$M^+ - SO_2$ (138)	0.12 (170°)	22 (140°)	1.4	60	0.5	0.02	0.24
Benzylalcohol (108)	$M^+ - HCO$ (79)	<0.02 (250°)	14 (250°)	<0.15	30	0.5	<0.005	<0.04

Diphenyldisulfide (218)							
$M^+ - HS$ (185)	<0.005 (190°)	4 (210°)	<0.6	11	0.10	<0.05	<0.05
$M^+ - 2S$ (154)	0.016 (190°)	6 (210°)	1.8	16	0.10	0.11	0.16
Dibenzylsultoxide (230)							
$M^+ - H_2SO$ (180)	<0.02 (160°)	11 (170°)	0.02	9	3.7	0.002	<0.005

Figure 5 shows the shape of the shifted peak $m/e = 120$ in the spectrum of valerophenone.

In all of these and of the following cases the shift of the rearrangement peak is interpreted as the effect of a small value of the maximum rate con-

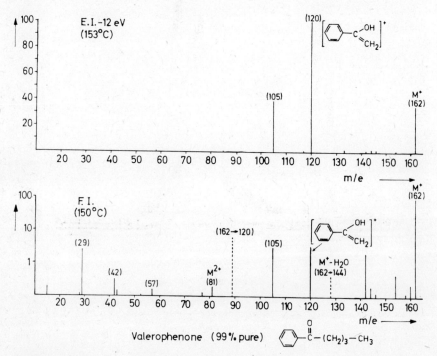

Figure 4 *EI* (12 eV) and *FI* mass spectra of valerophenone. (Corrected for ^{13}C-contributions. Note the linear and logarithmic intensity scales for *EI* and *FI*, respectively)

stant for the rearrangement reactions. In the sense discussed above, these rearrangements may be termed reactions with a low frequency factor. The frequency factor seems to decrease with the length of the alkyl chain in the series of ketones described here. In the *FI* spectrum of menthone, on the other hand, only a relatively strong metastable ion is observed at room temperature[7,9] indicating an even lower frequency factor.

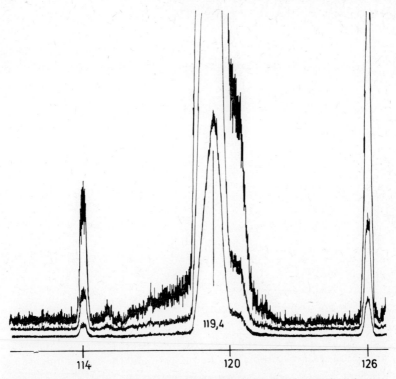

119,4

114   120   126

Figure 5 Shape of the shifted peak $m/e = 120$ originating from valerophenone. Octane ($m/e = 114.14$) and dimethylheptene ($m/e = 126.14$) additives for mass scale calibration. The peak tailing to lower masses is suppressed by the focusing system

II.1.2   *McLafferty rearrangements of aliphatic esters*

Three of the esters which we have studied will be discussed here: butanoic acid propyl-, butyl and amylester.

Three different types of rearrangements have been observed in the *EI* and *FI* spectra.

1) Double hydrogen transfer

2) McLafferty rearrangement on the alcoholic side

3) McLafferty rearrangement on the acid side

Both the first and the second rearrangement type are observed with relatively high relative intensities in the *FI* spectra, with shifted peak maxima. The rearrangement on the alcohol side, leading to an olefinic fragment ion, is even the base peak with the propyl and butylesters, and the second intense peak with the amylester. The double rearrangement is less intense. Finally the rearrangement on the acid side is only observed as a metastable ion.

The peak shifts towards smaller masses in the *FI* spectrum of the amyl ester amounts: ($U_{an} = +3$, $U_{cath} = -9$ kV).

Δm in mass units

$m/e = 89$	-0.87	± 0.08
$m/e = 70$	-0.25	± 0.08

Schulze et al.[10] have also observed intense rearrangement peaks in the *FI* spectra of esters. The $C_4H_8^+$ peak of n-butylacetate, for example, was found to be the base peak. Those authors applied the "metastable defocusing technique"[11-14] with a double focusing mass spectrometer in order to measure the energy distribution of the rearranged ions. From these measurements they also concluded that the rearrangement takes place in the gas phase with a maximum probability at about 10^{-11} sec after field ionization.

II.1.3 *McLafferty rearrangement of aromatic esters*

Very intense double rearrangement peaks at m/e 123 are found in the *EI* mass spectra of benzoic acid propyl- and butylester.

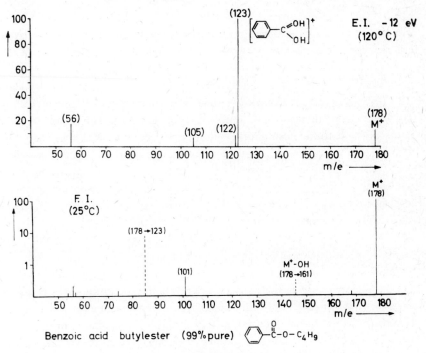

Figure 6 shows the *FI* and *EI* mass spectra of the last compound. In the *FI* mass spectra, on the other hand, this rearrangement peak can only be detected as an intensive metastable peak at room temperature.

If the emitter, however, is heated, a broad maximum between the theoretical mass position of the direct fragment ($m/e = 123$) and the metastable peak ($m/e = 84, 99$ for the butylester) is visible, with a maximum of intensity

Figure 6 *EI* (12 eV) and *FI* mass spectra of benzoic acid butylester (corrected for ^{13}C-contributions)

between $m/e = 100$–105 at an emitter temperature of $\sim 300\,°C$ ($U_{an} = 7,5\,kV$, $U_{cath} = 0$, lens system at earth potential). At higher temperatures the maximum increases in intensity and is more pronounced.

At the same time the maximum is shifted towards higher masses:

$$\text{at } t \sim 500\,°C \qquad \text{between} \quad m/e\ 111\text{–}113$$

$$\text{at } t \sim 800\text{–}900\,°C \quad \text{at} \qquad m/e\ 122.1.$$

Thus it can be concluded that the temperature raises the rate of reaction.

In the *FI* mass spectrum of phenylacetate a very strong rearrangement peak at $m/e = 104$ is observed which is the base peak in the spectrum.

$$m/e = 104$$

This reaction is a McLafferty rearrangement of the alcohol moiety via the ether oxygen. The peak is shifted by 0.20 ± 0.08 mass units.

II.1.4   Mass spectrometric ortho effect

This hydrogen rearrangement process will be discussed for the cases of salicylic acid methylester and 2-hydroxybenzylalcohol. The fast metastable peaks m_f near to the nominal mass number m are small with these two substances as compared with the normal metastable peaks m^* (at temperatures below $100\,°C$). However, the intensity of the m_f^* peak increases strongly with increasing temperature. Changing the temperature to $250\,°C$ increases the intensity of the $m/e = 106$ peak of 2-hydroxybenzylalcohol by a factor of 140. This strong rearrangement peak is found on the exact mass number $m = 106$ and is entirely sharp. A shifted peak was observed at about 0.5 mass units lower position if the *FI* emitter was activated with benzonitrile. This peak is broadened, in contrast to the sharp $m/e = 106$ peak (see Figure 7). The intensity of this shifted peak is much less temperature dependent than the sharp $m/e = 106$ peak. The different temperature dependence of the two peaks leads to the assumption of different mechanisms of origin of the peaks. Because of the sharpness of the $m/e = 106$ peak it can be assumed that it originates from a surface rearrangement reaction. The minimum between the direct and the shifted peak in Figure 7 can be due to a defocusing of the lens system.

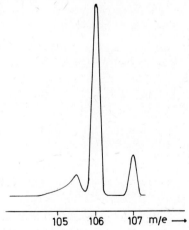

$m/e = 106$

There is also a direct bond fission leading to a fragment of $m/e = 107$. This peak is found at the normal mass position. The peak maxima originating from direct bond fissions are normally not shifted but found at the exact theoretical mass position, even if they have a long tail towards smaller mass numbers. A few exceptions have been found, however. A peak near $m/e = 105$ originating from a direct bond fission was found in the *FI* mass spectra of butyro-,valero- and caprophenone. The mass shift was about -0.2 mass units. In the *FI* spectrum of butanoic acid amylester two direct bond fission peaks near $m/e = 101$ and $m/e = 115$ were found. The peak shift could not be determined very accurately because of the small peak intensities, but the shift was of the order of -0.6 mass units. More work has to be done in order to give a well founded theoretical explanation for the assumption on that decreased maximum rate constants existed in these special cases.

Table 4 demonstrates the strong increase of the rearrangement peak of 2-hydroxybenzylalcohol with increasing temperature.

Figure 7 Part of the *FI* mass spectrum of 2-hydroxybenzylalcohol, showing the different shapes of the gas phase and surface rearrangement peak at $m/e = 105.5$ and $m/e = 106$

Table 4   Relative intensities of the rearrangement peak ($m/e = 106$), the direct bond fission peak ($m/e = 107$) and the normal metastable peak (m_n^*) of 2-Hydroxybenzylalcohol at two temperatures. (Intensities in % of the parent peak intensity)

	100°C	250°C
$m/e = 106$	0.06–0.1	8.3
$m/e = 107$	0.53	1.8
$m/e = \ 90.61$	0.26	2.5

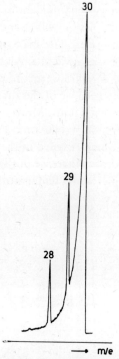

Figure 8   Typical peak shapes for direct bond fission. Note the sharp rise on the high mass side of the $m/e = 30$ peak ($CH_2NH_2^+$) of an amine

Summarizing one can state that hydrogen rearrangement peaks may be observed at their normal mass position, as fast metastable ions with shifted peak maximum or only as metastable ions. This demonstrates the strong dependence of the characteristic rate constants on the molecular structure. A characteristic rate is one which in connection with a given energy distribution leads to a peak intensity maximum.

II.2 Skeletal rearrangement recations

A skeletal rearrangement is defined by the equation

$$ABC \rightarrow AC + B, \quad AC \neq \text{hyorogen}$$

B frequently represents a neutral molecule. Figure 9 shows the *FI* and *EI* mass spectra of diphenyl sulfoxide.

The rearrangement processes by which SO or CO are eliminated, from the parent ion of diphenyl sulfoxide are found with a relative intensity of 27% and 8%, respectively, of the total fragment ion intensity in the 12 eV *EI* mass spectrum. These rearrangement fragments are completely absent in the *FI* mass spectrum (R and $S = 0$) within the limit of the measuring sensitivity, which leads to a lower limit of ion detection of 0.01% of the base peak. Metastable ions corresponding to the rearrangement processes M^+—SO and M^+—CO are formed with relative intensities of about 0.1%, this, indicating very slow decomposition processes.

The direct bond fission M^+—O, on the other hand, has a relatively high intensity in the *FI* spectrum, but shows no corresponding metastable peak. This indicates that the direct bond fission in this molecule is much faster than the rearrangement processes.

The elimination of SO from diphenyl sulfoxide is a relatively simple rearrangement. The elimination of CO from diphenylether is much more complicated, as indicated by the following mechanism:

The rearrangement peak $m/e = 142$ cannot be detected in the *FI* mass spectrum, whereas the corresponding metastable peak has a relative intensity of 0.5% (see Figure 10).

Table 3 shows the data for a number of skeletal rearrangement processes. The S and Q factors of all these rearrangement reactions are very low, as is to be expected for complicated processes which have a small frequency factor. Further, the activation energies of skeletal rearrangements are higher than those of hydrogen rearrangements which leads to a further reduction of the rate constants. The activation energies are smaller, on the other hand, than

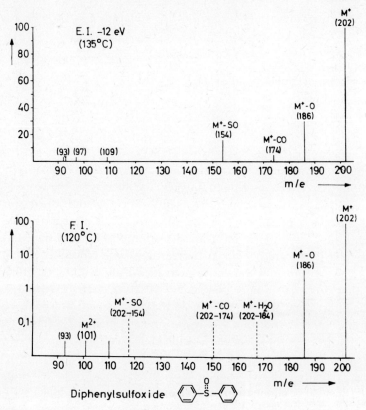

Figure 9 *EI* (12 eV) and *FI* mass spectra of diphenylsulfoxide (corrected for ^{13}C-contributions)

those of direct bond fissions of the same molecule. It can be concluded from the stereo-chemistry of these molecules that the surface does not induce a preferential rearrangement as compared to the gas phase. The possibility cannot be excluded, however, that some skeletal rearrangement processes with larger S and Q values could be detected in the future because no serious steric complications seem to be involved with some of these rearrangement reactions.

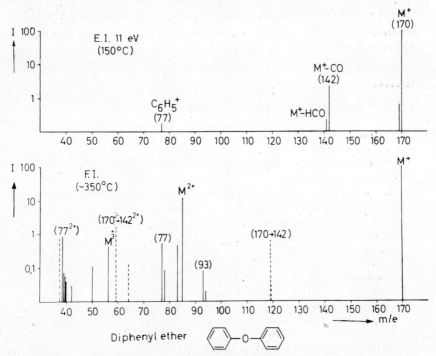

Figure 10   *EI* (11 eV) and *FI* mass spectra of diphenyl ether (corrected for ^{13}C-contributions)

Possibly some strongly shifted rearrangement peaks exist in the *FI* mass spectra of *p*-cresol, anthraquinone and dibenzyl sulfoxide. However, their intensity was found to be small and they could not yet be attributed to a specific process unambiguously. Therefore, they have not been included in Table 3.

Diminishing and expansion of rings has also been studied by the *FI* technique. This will be not discussed, however, in this article.

CONCLUSIONS

Summarizing, one can say that much new information on the kinetics and mechanisms of rearrangement reactions can be obtained by comparing low voltage EI with FI mass spectra. The new information can be extended if additional measurements on the distribution of rate constants are made with the

FI-peak shape analysis and retarding potential technique[7] and if reliable values for the activation energies of rearrangement reactions would be made by improved appearance potential measurements[15].

References

1. D.H.Williams and R.G.Cooks, *Chem. Comm.* 663 (1968).
2. R.G.Cooks, I.Howe, and D.H.Williams, *Organic Mass Spectrometry* **2**, 137 (1969).
3. H.M.Rosenstock, M.B.Wallenstein, A.L.Wahrhaftig, and H.Eyring, *Proc. Natl. Acad. Sci. USA* **38**, 667 (1952).
   H.M.Rosenstock and M.Krauss, *Adv. in Mass Spectrometry*, Vol.II, p.251, Pergamon Press, Oxford 1963.
   H.M.Rosenstock and M.Krauss, Chapter in *Mass Spectrometry of Organic Ions* (F.W.McLafferty Ed.), Academic Press, New York 1963.
4. O.Osberghaus and Ch.Ottinger, *Phys. Letters*, **16**, 121 (1965).
5. Ch.Ottinger, *Z. Naturforsch.* **22a**, 20 (1967).
6. I.Hertel and Ch.Ottinger, *Z. Naturforsch.* **22a**, 1141 (1967).
7. H.D.Beckey, H.Hey, K.Levsen and G.Tenschert, *J. Mass Spectry. and Ion Phys.* **2**, 101 (1969).
8. H.D.Beckey and H.Knöppel, *Z. Naturforsch.* **21a**, 1920 (1966).
9. H.D.Beckey and H.Hey, *Org. Mass Spectry.* **1**, 47 (1968).
10. P.Schulze, W.J.Richter, and A.L.Burlingame, *17th Ann. Conf. on Mass Spectrometry*, Committee ASTM, E-14, Dallas (1969).
11. M.Barber and R.M.Elliott, *12th Ann. Conf. on Mass Spectrometry*, Committee ASTM, E-14, Montreal, 1964.
12. J.H.Beynon, R.A.Saunders, and A.E.Williams, *Nature*, **204**, 67 (1964).
13. J.H.Futrell, K.R.Ryan, and L.W.Sieck, *J. Chem. Phys.* **43**, 1832 (1965).
14. K.R.Jennings, *J. Chem. Phys.*, **43**, 4176 (1965).
15. P.Potzinger and G.v.Bünau, *Ber. d. Bunsenges.*, in press (1969).
16. E.N.Chait, F.W.Shannon, W.O.Perry, G.E. van Lear and F.W.McLafferty, *Int. J. Mass Spectry. and Ion Phys.* **2**, 141 (1969).

Techniques of combined gas chromatography and mass spectrometry

C. MERRITT, JR.

U.S. Army Natick Laboratories, Natick, Massachusetts

The techniques of gas chromatography and mass spectrometry are ideally suited to be combined in a method for the analysis of complex mixtures which provides in a single system for the separation of the components, their identification and quantitative determination. This analysis system has grown from its crude inception over a decade ago[1,2] to an elegance hardly matched in any other analytical procedure. Nearly all modern mass spectrometers have a fast scan capability and a means of attaching a gas chromatographic inlet. The operational parameters of both techniques share many common factors. Separation in the gas chromatograph and analysis in the mass spectrometer both occur in the vapour state. The temperature range of compounds analysed and the sample size accommodated are comparable for both. Preparation and manipulation of samples prior to analysis is usually the same for either gas chromatography or mass spectrometry and little, if any, modification is required when the techniques are combined. One inherent incompatibility, however, has existed between the two methods, namely, the requirement for reduced pressure in the spectrometer ion source and the relatively large volumes of carrier gas needed to elute the chromatography column. The solutions to this problem have been many and varied, depending on the type of chromatography column used, the kind of mass spectrometer employed, and nature of the mixtures to be separated and analysed.

In a recent cursory survey of the literature more than 50 papers were found describing coupled gas chromatograph-mass spectrometer systems. Some recent reviews[3-6], however, provide the reader with an overview of the subject and an excellent summary of the current state of the art. The treatment of combined gas chromatography-mass spectrometry systems in previous editions in this series[7,8] has been concerned with pertinent aspects and new developments rather than overall review. This tradition will be emulated here, and the paper will be limited to consideration of parameters associated with the combined system.

ION SOURCE PRESSURE

The principal requirement which dictates the method of coupling the gas chromatograph to the mass spectrometer is the pressure at which the ion source must operate. With batch inlet systems, spectra may be obtained from 10^{-10} to 10^{-8} gms of sample and ion source pressures are usually less than 10^{-6} Torr. When a GC inlet is used, the same amount of sample is required, but the components are diluted from 100 to 10,000 fold with helium, depending on the type of column used. Source pressure from GC inlets therefore is primarily due to the carrier gas and the partial pressure due to the components to be analysed is very small. The question arises, then, as to how to best accommodate the large carrier gas volume to the spectrometer.

In resolving the dichotomy of the gas flow and pressure requirements of the gas chromatograph and the mass spectrometer two main approaches have been developed, both of which have received rather wide acceptance in recent years. One method utilizes the low flow rate characteristics of small open tubular columns. The other, used mainly with packed columns, employs some type of molecular separator to split off the bulk of the carrier gas stream and to concentrate the eluates in the effluent reaching the spectrometer. In both instances the objective is to reduce the pressure in the ion source to what may be considered a desirable range.

Many factors contribute to a consideration of what is a suitable ion source pressure. First, of course, is the quality of the spectra obtained, but in addition, operational factors, such as maintenance of source stability, prevention of multiplier contamination, the life of the filament, and freedom from high voltage arcs are all matters of concern. In general, source pressures in the range below 5×10^{-6} Torr have been considered desirable for most types of mass spectrometers when coupled to a gas chromatograph. The pressure achieved, however, depends not only on the carrier gas flow in the chromatographic column and, correspondingly, the rate at which it enters the spectro-meter from the splitter, separator, restrictor or whatever interfacing device may be employed, but also on the characteristics of spectrometer vacuum system.

The low flow rates employed in open tubular columns can be easily accommodated by the pumping speeds used in some modern mass spectro-meters, but the higher flow rates used in packed columns require greater pump capacities to maintain the source pressure in a desired range. Unfortunately, very few mass spectrometers commercially available today have vacuum systems that provide the needed pumping capacities to handle GC effluents in a dynamic way. The combined system, therefore, in most instances must be designed to overcome this constraint, hence, the widespread use of splitters and separator devices to reduce the carrier gas volume entering the spectrometer. A few instruments, of the quadrupole or time of flight type, are equipped with vacuum systems which permit a GC inlet to be directly coupled. In many cases, particularly with the magnetic scanning instruments, an adequate pumping system is throttled by the use of cold traps and valves that reduce the overall conductance of the system. It is to be fervently hoped that future designers of mass spectro-meters will take a broader view of the instrumentation and provide larger vacuum systems that will allow the spectrometer a greater scope in its ac-commodation of inlet devices.

Although operation of the spectrometer at ion source pressures less than 10^{-6} is normally considered desirable, the effect of higher helium pressures in the ion source upon the spectrum has been studied but little. Teeter, *et al.*[9], and Leemans and McCloskey[10] have reported a decrease in the relative abundance of the molecular ion in the spectra of certain higher molecular weight compounds as the helium pressure increased. In some circumstances there also appears to be a broadening of the spectrum peaks[10,11] with a re-sulting loss in resolution, but the effect appears to be due to disturbances in the analyser or detector regions of the spectrometer and would not be ob-served in instruments employing differential pumping of the ion source, or a means of helium ion defocusing. Brubaker[12] has recently reported the opera-tion of a quadrupole mass spectrometer with a GC inlet employing hydrogen as the carrier gas with no loss of resolution or sensitivity at pressures up to 1×10^{-4} Torr.

Because of the inability of many spectrometers to remove the carrier gas, the small amounts of sample in eluates from open tabular columns and the poor recovery ratios from separators used with packed columns, it would be advantageous to operate a spectrometer ion source at higher pressures if no adverse effects are produced.

The effect of increased carrier gas pressure on the spectrum has been studied on three different types of mass spectrometer[13]. The instruments used in the investigation were a CEC Model No. 21–110B double focusing high resolution mass spectrometer, a Bendix Model 14 time of flight mass spectrometer and an EAI Model 300 quadrupole mass spectrometer. The ion source of each was fitted with a molecular leak inlet to allow the compound whose spectrum was to be observed to be admitted at a constant rate. A separate inlet (actually the normal GC inlet port) was used to admit helium at appropriately chosen flow rates to produce the desired pressure in the ion source. In all cases helium ions were defocused from the detector. Predynode gating at the multiplier was used in the time of flight instrument. In the quadrupole, the *rf* sweep was started above mass 10. In the double focusing instrument, the electron beam collimating magnets provide a field which deflects helium ions from the focus slits. The ion source of the high resolution instrument was isolated from the electrostatic analyser by the focus slits and differentially pumped. The spectrum was displayed on an oscilloscope for purposes of monitoring the performance of the spectrometer and to provide photographs of the spectrum from which to prepare Figure 1. Spectra were also recorded on an oscillograph to permit accurate measurement of mass number and abundance ratios as well as relative sensitivity values. These data are given in detail elsewhere[13]. The results of this study are summarized in Figure 1, which shows that in the three spectrometers employed the spectrum is essentially unaltered by the presence of helium in the pressure range from 2×10^{-6} to 5×10^{-5} Torr. In all cases neither pattern nor resolution was changed. The sensitivity was found to decrease in the time of flight spectrometer as pressure increased, and in the quadrupole, it was found to increase in the range from 3×10^{-5} to $\sim 7 \times 10^{-5}$. Neither effect, however, was observed when the spectra were produced at low ionizing voltage (18 eV). The phenomena are apparently associated with helium ions rather than helium atoms. The sensitivity increase is perhaps due to secondary ionization behavior induced by the helium, whereas the decrease may be due to poor ion transmission. In any event, the variations in sensitivity do not appear randomly, are easily observed, and would not interfere with the normal acquisition of spectra from a chromatographic eluate.

The effect of helium pressure in the ion source was also studied at high resolution. Typical results are shown in Figure 2. The doublet at mass 103 from a mixture of styrene and benzonitrile is separated by less than a 10% valley at a resolving power of about 1 part in 8000.

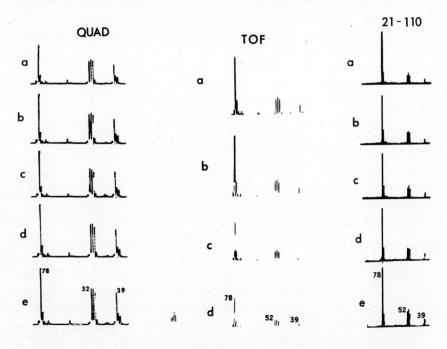

Figure 1 Photographs of oscilloscope presentation of selected mass spectra of benzene shown at various ion source pressures of helium carrier gas on certain spectrometers

QUAD: EAI Model 300 quadrupole mass spectrometer; a) 2×10^{-6} Torr, b) 5×10^{-6} Torr, c) 7×10^{-6} Torr, d) 2×10^{-5} Torr, e) 5×10^{-5} Torr. TOF: Bendix Model 14 time of flight mass spectrometer; a) 2×10^{-6} Torr, b) 5×10^{-6} Torr, c) 1×10^{-5} Torr, d) 5×10^{-5} Torr. 21–110: CEC Model 21–110 double focusing high resolution mass spectrometer: a) 2×10^{-6} Torr, b) 4×10^{-6} Torr, c) 7×10^{-6} Torr, d) 1×10^{-5} Torr, e) 3×10^{-5} Torr [m/e 78 peak off scale in all spectra]

The knowledge that usable spectra of components in GC effluents may be obtained at higher ion source pressures than usually employed greatly relieves the limitations frequently invoked in coupling the column to the spectrometer.

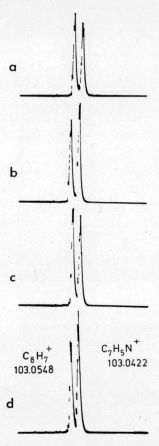

Figure 2   Photographs of oscilloscope presentation of a portion of the mass spectrum of a mixture of styrene and benzonitrile showing the resolution of the doublet and at m/e 103 $[\Delta m/m \sim 8200]$

a) 3×10^{-6} Torr,   b) 8×10^{-6} Torr,   c) 1×10^{-5} Torr,   d) 5×10^{-5} Torr

GC–MS INTERFACES

Historically, three types of interface, or a combination thereof, have been used to connect the chromatograph column to the spectrometer. A direct inlet, with or without a restrictor, a stream splitter, or a molecular separator may all be used depending on the particular characteristics of the column and the spectrometer.

With the early systems, however, and to some extent even today, mass spectroscopists have seemed content to try to adapt the operation of the mass spectrometer to the effluent of the gas chromatograph without seeking to find variations in gas chromatographic parameters that might be more suitable to the constraints of the spectrometer. In general, it may be said that a greater variation can be found in the mode of performing gas chromatographic separations, than in the conduct of a mass spectrometric analysis. Traditionally, packed columns, necessarily employing high carrier gas flow rates, have been more widely used than the open tubular type of column. Recent studies[14] suggest that the use of small bore open tubular columns possess operational characteristics that are more compatible in a combined GC/MS system.

When packed columns are employed, however, it is necessary to divert a portion of the carrier gas stream in order to reduce the pressure of the ion source to within operational limits. For most mass spectrometers this pressure is 10^{-4} Torr or less. For optimum performance, particularly in magnetic scanning mass spectrometers, the pressure should be much lower.

The adjustment of carrier gas flow rate to produce the desired ion source pressure was accomplished in some of the early systems[1,2,15-18] by a simple stream splitter (usually a tee and a needle valve) which could be regulated to remove the excess gas stream. With $\frac{1}{4}$ in. columns it was necessary to vent more than 99% of the gas chromatographic effluent. The ratio was more favourable with $\frac{1}{8}$ in. columns which allowed about 10% of the eluate to enter the ion source, but the loss of substantial amounts of the eluents in either case reduced the overall sensitivity of the method and many of the components present in small amounts were undetected.

The development of molecular separators contributed greatly to the utilization of packed columns in G.C. inlets for mass spectrometers. Basically, three types of separtors have been employed. They are depicted schematically in Figure 3. In all, the mode of operation is to eliminate the helium preferentially, thus concentrating the eluents in the gas stream entering the spectrometer. In the Becker–Ryhage type[19-22], (Figure 3a), the gas stream enters an evacuated chamber through a jet. The lighter helium atoms tend to diffuse outward into the area of reduced pressure more rapidly than the heavier organic molecules. The enriched stream tends to flow toward the sample inlet of the mass spectrometer. The Watson–Biemann type[23,24] (Figure 3b) is based on the difference in rate of molecular diffusion through a porous medium such as a glass frit[23,24], or Teflon[25] or metal[26,27,28] membranes. In the Llewellyn[29] type (Figure 3c) the relatively greater permeability of a polymer membrane to organic molecules than to helium is used to accomplish the enrichment. The performance of all three types may be improved by employing a second stage[29,30,32].

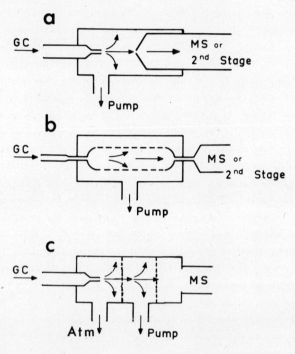

Figure 3   Schematic diagrams of various molecular separators:
a) jet type, b) porous membrane type, c) permaselective membrane type

There appears to be considerable variation in the performance of molecular separators. The enrichment (i.e. relative increase in component concentration to helium) and yield (i.e. the percentage of the total amount of component reaching the ion source) of the devices varies with geometry of construction, temperature, carrier flow rate, and the pumping speeds of the separator and spectrometer vacuum systems. Performance has been evaluated by the originators[21,24,25,26,30] and by several users[31,32,33,34,35]. In general, although the separator allows the mass spectrometer to be operated at lower source pressures, the recovery ratios are poor, and in spite of the enrichment factor, the presence of small amounts of components, especially in unknown samples, may go undetected, or the mass spectrum may not be sufficiently intense to permit identification.

Another difficulty arising with the use of separators in the analysis of mixtures having a wide range of molecular weights is a change in the distribution of components in the gas stream entering the spectrometer. In the case of the diffusion type separators, the problem is basic since rate of diffusion is a function of molecular weight. With the permeable membrane type, discrimination is due to selective adsorption. This may perhaps be overcome by appropriate selection of the membrane material, but it is difficult to conceive of a substance having the property of equal permeability to all organic substances, polar and nonpolar alike, over a range of molecular sizes. Moreover, the two stage membrane separator has been shown to exhibit considerable band spreading[35] as the eluent moves across the interface from chromatograph to spectrometer. In addition, to band spreading and distortion of composition all the separators have on various occasions been reported to contribute to other artefacts in the analysis. Either components are removed by adsorption on the separators[32] or decomposed by thermal cracking induced on their metal or glass surfaces[36].

Separators have also been employed with capillary columns[22,24,31,32,34,35] but there is little justification except in special circumstances. Since carrier gas flow is usually very low ($\sim$1–5 ml/min) in open tubular columns, the enrichtsment achieved is only marginal.

Many investigators have successfully coupled 0.01 to 0.03 in. open tubular columns to a mass spectrometer by means of a simple direct inlet[14,37,38,39] or, in cases where pressure limitations in the spectrometer made it necessary, through a simple stream splitter[16,17,18,40]. Two problems have been encountered. The smaller capillary columns are restricted to small sample sizes, whereas the larger columns failed to provide good efficiency when a high pressure drop was induced through the column from the spectrometer vacuum.

The requirements for gas chromatographic separation are determined by the composition of the very complex mixtures which are frequently encountered. For example, the major components of such mixtures are usually present in quantities of the order of a few micrograms, with a range of component concentration that may vary by more than 2000 to 1. In addition, many peaks do not represent single components, but mixtures of 2, 3 or more compounds. In order to accommodate such samples two conditions must be achieved. The largest sample size must be available to the spectrometer, while maintaining the greatest possible attainable separation. In general, only open tubular columns of high capacity will meet these requirements. Accordingly, it was decided to investigate in detail the properties of this type of column and its compatibility with a mass spectrometer.

Optimum G.C. column performance imposes rather strict limitations on the flow of carrier gas[41,42]. Likewise, however, the amount of carrier entering the ion source must be controlled for good spectrometer behaviour. Ideally then, in a tandem configuration of a gas chromatograph with a mass spectrometer, the desirable operating characteristics of each system should be preserved. Thus, if the drop in pressure from the column outlet into the spectrometer ion source can be made to occur between these two points, the desired result will be achieved. This may be accomplished by the appropriate design of a pressure dropping restrictor between the column and the ion source. A diagram of one type of restrictor which is found to be suitable and easily constructed is shown in Figure 4. A length of 0.006 inch capillary tubing is connected to a vacuum manifold and is cut to a length which for a pressure of 1 atmosphere of helium on the outside gives a selected pressure in the range of 1 to 5×10^{-5} on the inside. The section is then silver-soldered into sections of two larger diameter tubes so that the restrictor can be accommodated into a standard $\frac{1}{8}''$ Swagelock fitting for connection between the column and the spectrometer. This device is about 7″ long. Other restrictors can be constructed in a similar way from glass—or by merely crimping a piece of ordinary 0.01″ capillary tubing. Needle valves as part of a stream splitter have also been used as a restrictor[43] but in the author's laboratory they have been found to lack the precision needed to produce the required pressure drop for open tabular columns. Moreover, valve components do not

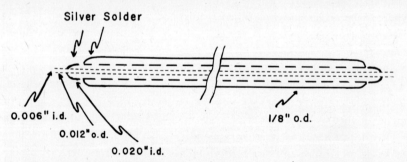

Figure 4 Schematic diagram of pressure dropping restrictor used for direct coupling of gas chromatography column to mass spectrometer ion source

function well at higher temperatures, and after repeated use the needle tends to stick easily in the closed position making the valve useless. Such valves are also frequently found to leak badly, thus contributing a high background spectrum of air.

There is nothing novel or original in the use of a restrictor between the column and the spectrometer—for many workers[11,14,15,17,18,31,37,38,40] have employed valves or capillary restrictors in a variety of systems that have been described—and they are frequently used in connection with splitters or separators to control the ratio of gas entering the spectrometer relative to the amount vented by the splitter. In all accounts reported in the literature, however, the coupling of open tubular columns has been directly into the spectrometer vacuum, so that the chromatography column is operating with the outlet below atmospheric pressure. Although Teranishi[44] has reported that there appears to be no deterioration of column performance under these circumstances, (at least for 0.01 capillaries), with larger bore columns there appear to be several difficulties. In addition to losses in column efficiency, the stationary phase may bleed excessively and leaks in the system may lead to a large background spectrum in the spectrometer.

If the restrictor is constructed to allow the column outlet to remain at atmospheric pressure, optimum column performance can be achieved. This is illustrated in Figure 5.

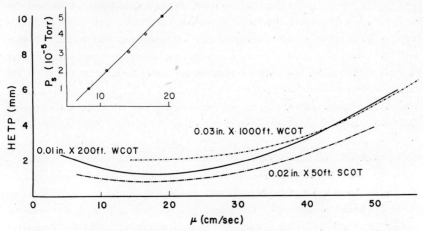

Figure 5 Graph showing relationship of chromatographic column efficiency (HETP) to carrier gas velocity for various type open tubular columns. Inset: Relationship of ion source pressure to carrier gas velocity. (EAI Model 300 quadrupole mass spectrometer)

Van Deemter plots for several open tubular type columns such as a conventional $0.01'' \times 200$ ft wall coated capillary, a larger bore $0.03'' \times 1000$ ft wall coated capillary and a $0.2'' \times 50$ ft support coated open tubular column are all shown. The range for optimum linear carrier gas velocity is seen in all cases to be about 15 to 20 cm/sec. In addition, the efficiency/flow characteristics for $0.02'' \times 500$ ft wall coated capillary and a 50 ft $\times 0.04''$ multichannel column[45] have been evaluated. These also have optimum linear velocities in the same range. The results of a study of ion source pressure as a function of flow rate through the restrictor is shown in the inset to Figure 5. The pressure remained in the range from 1 to 5×10^{-5} for carrier gas velocities between 25 to 20 cm/sec. These pressures, as seen in the preceding section, are entirely suitable for good mass spectrometer performance. Accordingly, it would appear that the range of flow rate for optimum column performance of open tubular type columns coincides ideally with suitable spectrometer ion source pressures when the column is isolated by a pressure dropping restrictor. In this mode of operation the column outlet pressure is about atmospheric and the column behaves as if no spectrometer were there.

An example of the performance of the restrictor is shown in Figure 6. The spectrometer total ion current chromatogram of the separation of the components of a sample of so-called high purity nanograde hexane on a 0.02 in. $\times 50$ ft squalane support coated open tubular column is shown at the top left. At the lower left, a thermal conductivity trace is shown. This was obtained from a microvolume thermal conductivity cell whose sensor element is arranged in series with the column. The katherometer was used to provide a chromatographic output which is separate from the spectrometer and permits the evaluation of such parameters as linear carrier velocity, retention times and peak shape independently of the effect of spectrometer vacuum. The thermal conductivity signal was recorded along with the spectrometer total ion current signal on separate channels of a dual channel potentiometer recorder. No band spreading occurs as the components pass through the restrictor from atmospheric pressure to ion source vacua. It was interesting to note that the microvolume thermal conductivity cell could be operated with a sensitivity that was comparable to the ion current monitors and a flame ionization detector. Although frequently overlooked, the thermal conductivity cell provides a useful means of monitoring the GC eluate in systems where a split to an FID is undesirable and where the TIC monitor will not operate while spectra are being scanned (as with most double focusing magnetic scanning mass spectrometers).

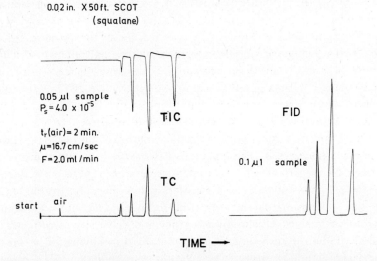

Figure 6 Chromatograms of a mixture of hexane isomers obtained with various detectors. TIC: Total ion current monitor of EAI Mode 300 quadrupole mass spectrometer. TC: Gow-Mac Model 10–952 Hex-nano katherometer (sensor inserted between column and restrictor). FID: Barber–Colman Series 5000 gas chromatograph with flame ionization detector (separate sample)

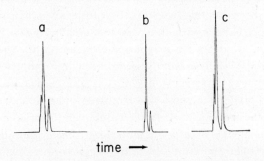

Figure 7 Comparison of the separation of hexane isomers on the same SCOT column, a) in a Barber–Colman series 5000 modular gas chromatograph with FID detector, b) coupled to a Perkin–Elmer Model 270 mass spectrometer through a glass frit separator, and c) coupled directly to an EAI Model 300 quadrupole mass spectrometer. Chromatograms in b) and c) are recordings of spectrometer total ion current. Column: 50 ft. × 0.02 in. Carbowax 1540; temperature, ambient; sample size, 0.05 μl; flow rates, a) 2 ml/min, b) 8 ml/min, c) 2 ml/min; ion source pressure, b) 5×10^{-7} Torr, c) 1×10^{-5} Torr [reprinted from ref. 14]

The chromatogram on the right of Figure 6 is shown to provide a comparison of the performance of the same column in a conventional gas chromatograph. The sample size is twice as large which accounts for the increased size of the peaks, while the slightly poorer resolution and efficiency can be attributed to a less efficient injector and a somewhat higher column temperature in the gas chromatograph used.

Figure 7 shows a set of chromatograms for the separation of the same mixture as shown in Figure 6. Unfortunately, these comparisons were made with a 0.02″ × 50 ft C.W. 1540 support coated open tubular column so the resolution is much poorer. However, they serve to illustrate the effect of using a separator. On the left is the FID chromatogram for the separation performed when the column is installed in a conventional gas chromatograph. On the right, the TIC chromatogram from the direct restrictor coupled G.C.–mass spectrometer system. Resolution, efficiency and sensitivity of response are about the same for corresponding sample sizes and flow rates on the gas chromatograph and the direct coupled GC–MS system. When a molecular separator is used, as seen in the centre chromatogram, the ion source pressure could be reduced to the 10^{-7} range, but column flow rate had to be increased from 2 ml/min to at least 8 ml/min to achieve the same separation efficiency. When the flow rate is thus increased the column is being operated at less than optimum linear velocity for minimum HETP. In addition, the sample size had to be increased 5 fold to account for the loss in this splitter. In most cases therefore, there appears to be little advantage in thus employing a molecular separator since the spectra obtained in the range of 10^{-5} Torr in the ion source were equally as good as those in the 10^{-7} range.

Figure 8 shows a separation of a somewhat more complex sample on the direct coupled GC/MS system. The sample is a mixture of alkanes separated on a 0.02 in. × 50 ft squalane support coated open tubular column. This chromatogram was obtained when the column and restrictor were attached to an EAI Model 300 Quadrupole mass spectrometer, but identical results were obtained when the same separation was performed with this column and inlet attached to a Bendix Model 14 time of flight mass spectrometer or a CEC Model 21–110B high resolution mass spectrometer. Similar results have been obtained with other types of open tubular columns, in particular, with 0.02 and 0.03 in. columns.

The 0.02 in. support coated open tubular columns and the 0.03 in. wall coated columns have sample capacities which approach those of $\frac{1}{8}$ in. packed columns. With this arsenal of open tubular columns all capable of

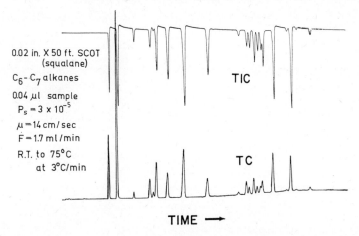

0.02 in. X 50 ft. SCOT (squalane)

$C_6 - C_7$ alkanes

0.04 μl sample

$P_s = 3 \times 10^{-5}$

$\mu = 14$ cm/sec

$\dot{F} = 1.7$ ml/min

R.T. to 75°C at 3°C/min

TIC

TC

TIME →

Figure 8 Simultaneous chromatograms of a mixture of alkanes obtained with dual detectors. TIC: total ion current monitor of EAI Model 300 quadrupole mass spectrometer. TC: Gow–Mac Model 10–952 Hex-nano katherometer

being directly coupled to the mass spectrometer without a splitter and operating at optimum efficiency, it now appears that the investigator of complex mixtures can have both the desired separability and capacity in his selection of columns.

References

1. J.C. Holmes and F.A. Morrell, *Appl. Spec.*, **11**, 86 (1957).
2. R.S. Gohlke, *Anal. Chem.*, **31**, 535 (1959).
3. W.H. McFadden, *Advances in Chromatography*, Vol. **4**, J.C. Giddings and R.A. Keller, eds., Marcel Dekker, Inc., New York, 1967, pp. 265–332.
4. W.H. McFadden, *Separation Science*, **1**, 723 (1966); see also *Separation Techniques in Chemistry and Biochemistry*, R.A. Keller, Ed., Marcel Dekker, Inc., New York, 1967, pp. 263–286.
5. C. Merritt, Jr., *Applied Spectroscopy Reviews*, in press.
6. A.B. Littlewood, *Chromatographia*, **1**, 40 (1968).
7. R.I. Reed, ed., *Mass Spectrometry*, Academic Press, London, 1964.
8. R.I. Reed, ed., *Modern Aspects of Mass Spectrometry*, Plenum Press, New York, 1968.
9. R.M. Teeter, C.F. Spencer, J.W. Green, and L.H. Smithson, *J. Am. Oil Chem. Soc.*, **43**, 82 (1966).
10. F.A.J.M. Leemans and J.A. McCloskey, *J. Am. Oil Chem. Soc.*, **44**, 11 (1967).
11. H. Widmer and T. Gaumann, *Helv. Chim. Acta*, **45**, 2175 (1962).
12. W.M. Brubaker and W.S. Chamberlain, Paper No. 79 presented before the *17th Annual Conference on Mass Spectrometry and Allied Topics*, ASTM Committee E-14, Dallas, Texas, May, 1969.

13. C.Merritt, Jr., M.L.Bazinet, and W.G. Yeomans, *Chemical Instrumentation*, in press.
14. C.Merritt, M.L.Bazinet, and W.G.Yeomans, *J. Chromatog. Sci.*, **7**, 122 (1969); see also *Advances in Gas Chromatography—1969*, A.Zlatkis, ed., Preston Technical Abstracts Co., Evanston, Illinios, 1969, pp. 209–210.
15. E.Selke, C.R.Scholfield, C.Evans, and H.J.Dutton, *J. Am. Oil Chem. Soc.*, **38**, 614 (1961).
16. D.Henneberg and G.Schomburg, *Gas Chromatography*, **1962**, M. Van Swaay, ed., Butterworths, London, 1963, pp. 191–203.
17. J.A.Dorsey, R.H.Hunt, and M.J.O'Neal, *Anal. Chem.*, **35**, 511 (1963).
18. W.H.McFadden, R.Teranishi, D.R.Black, and J.C.Day, *J. Food Sci.*, **28**, 316 (1963).
19. E.W.Becker, *Separation of Isotopes*, H.London, ed., George Newnes Ltd., London, 1961, Chapter 9.
20. E.Stenhagen, *Z. Anal. Chem.*, **205**, 109 (1964).
21. R.Ryhage, *Anal. Chem.*, **36**, 759 (1964).
22. R.Ryhage, S.Wikstrom, and G.R.Waller, *Anal. Chem.*, **37**, 435 (1965).
23. J.T.Watson and K.Biemann, *Anal. Chem.*, **36**, 1135 (1964).
24. J.T.Watson and K.Biemann, *Anal. Chem.*, **37**, 844 (1965).
25. S.R.Lipsky, C.G.Horvath, and W.J.McMurray, *Anal. Chem.*, **38**, 1585 (1966).
26. R.F.Cree, *Pittsburgh Conference Analytical Chemistry and Applied Spectroscopy*, March, 1967.
27. M.Blumer, *Anal. Chem.*, **40**, 1590 (1968).
28. D.P.Lucero and F.C.Haley, *J. Gas Chromatog.*, **6**, 477 (1968).
29. P.M.Llewellyn and D.P.Littlejohn, *Pittsburgh Conference on Analytical Chemistry and Applied Spectroscopy*, February, 1966.
30. R.Ryhage, *Arkiv. Kemi.*, **26**, 305 (1966).
31. J.A.Vollmin, I.Omura, J.Seibl, K.Grob, and W.Simon, *Helv. Chim. Acta*, **49**, 1768 (1966).
32. M.Ct.N. de Brauw and C.Brunnee, *Z. Anal. Chem.*, **229**, 321 (1967).
33. M.A.Grayson and C.J.Wolf, *Anal. Chem.*, **39**, 1438 (1967).
34. W.D.MacLeod, Jr., *J. Gas Chromatog.*, **6**, 591 (1968).
35. D.R.Black, R.A.Flath, and R.Teranishi, *J. Chromatog. Sci.*, **7**, 284 (1969); see also *Advances in Chromatography—1969*, A.Zlatkis, ed., Preston Technical Abstracts Co., Evanston, Illinois, 1969, pp. 203–208.
36. R.Ryhage and E. von Sydow, *Acta Chem. Scandinavica*, **17**, 2025 (1963).
37. D.Henneberg and G.Schomburg, *Advances in Mass Spectrometry*, **Vol. 4**, E.Kendrick, ed., Elsevier, London, 1968, pp. 333–343.
38. G.Schomburg and D.Henneberg, *Chromatographia*, **1**, 23 (1968).
39. R.S.Gohlke, *Anal. Chem.*, **34**, 1332 (1962).
40. A.E.Banner, R.M.Elliott, and W.Kelly, *Gas Chromatography—1964*, A.Goldup, ed., Elsevier, London, 1965, pp. 180–189.
41. H.Purnell, *Gas Chromatography*, John Wiley & Sons, New York, 1962. S. Dal Nogare and R.S.Juvet, *Gas-Liquid Chromatography*, Interscience, New York, 1962.
42. J.C.Giddings, *Dynamics of Chromatography—Parts I & II*, Marcel Dekker, Inc., New York, 1965.

43. F.W.Karasek, *Research/Development*, **20**, 28 (1969).
44. R.Teranishi, R.G.Buttery, W.H.McFadden, T.R.Mon, and J.Wasserman, *Anal. Chem.*, **36**, 1509 (1964).
45. J.T.Walsh and C.Merritt, Jr., *J. Gas Chromatog.*, **5**, 420 (1967); see also *Advances in Gas Chromatography—1967*, A.Zlatkis, ed., Preston Technical Abstracts Co., Evanston, Ill., 1967, pp. 48–51.

Nuclear measurements by mass spectrometry

N. R. DALY and N. J. FREEMAN

United Kingdom Atomic Energy Authority, Aldermaston, England

The mass spectrometer is an instrument that has developed considerably over the last fifty years, and the reason for this is to be found in the ever-enlarging new areas of science and technology to which mass spectrometry can be applied. Isotope separators have developed over much the same period and since the techniques have a lot in common, they will also be considered in this paper.

1 INTRODUCTION

Thomson[1] in 1912 found strong evidence for the existence of isotopes. The confirmation of their existence by Aston[2] in 1919 was one of the fundamental advances that led to the present understanding of the atomic nucleus. Since the time that Aston's machine was developed there have been many advances in the instrumentation side, and a large field of application has arisen for mass spectrometers. Widely differing problems, such as measuring neutron cross sections of atoms, examining the minute quantities of gases present in meteorites, or measuring burn up in nuclear reactors can be tackled by modern instruments. Specialised machines have been developed as the field of applications has widened.

In 1934 the first isotope separator to separate appreciable quantities was constructed at the Cavendish Laboratory by Oliphant *et al.*[3]. Developments in America led to the construction of the Calutrons for the separation of gramme quantities of isotopes, whereas in Europe designers concentrated mainly on the smaller high enrichment machines described later.

213

It is the purpose of this paper to review some of the types of machines and the variety of nuclear physics problems to which they can be applied.

Almost all mass spectrometers and isotope separators consist of the following four basic elements:

1   an ion source that ionizes and accelerates the solid, liquid or gaseous material under examination;

2   a magnetic analyser that deflects ions into trajectories which depend on their mass to charge ratio;

3   an ion detector that records the ion beams or, in the case of a separator, plate to collect the isotopes;

4   a vacuum system to reduce the pressure in the flight path of the ion beams to between 10^{-6} and 10^{-10} Torr.

A sketch is shown in Figure 1 illustrating the principles of Aston's im proved mass spectrograph. He produced positive ions in a discharge tube and these possessed a wide range of energies. Beam collimation was achieved by using two slits S_1 and S_2 and the beam was deflected by a uniform electric field between parallel plates. A magnetic field directed into the plane of the paper deflected and focused the beam onto a photographic plate where an observable image was formed. Aston's simple instrument possessed velocity focusing whereby ions with the same e/m, where e and m are the charge and mass respectively, focused at the same point although having an energy distribution from the discharge tube ion source. No direction focusing was present in Aston's design, but the difficulty was overcome by strong beam collimation.

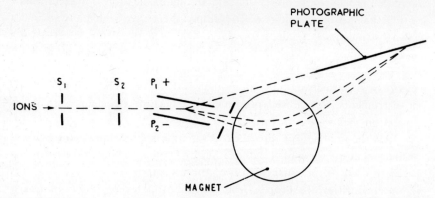

Figure 1   Aston's mass spectrograph

When ions are formed with small kinetic energy a velocity analyser is not required and the mass spectrometer becomes, in principle, simpler than Aston's machine. The discovery by Stephens[4] in 1934 of the focusing properties of sector-type magnetic fields led to the geometry of many machines in use today. A typical single stage instrument is shown in Figure 2. Ion beam focusing is achieved with uniform magnetic fields when the source slit, magnet "apex", and detector slit are collinear. The deflexion of ions in this system is governed by the well known mass spectrometer equation:

$$\frac{m}{e} = \frac{r^2 H^2}{2V}$$

where H is the magnetic field, r the radius of curvature, and V the accelerating potential. Scanning in single stage instruments can be carried out by variation of either the magnetic field or the accelerating potential. Voltage scanning is not preferred since it varies the ion source efficiency, an effect that must be allowed for in some measurements.

The resolution of a mass spectrometer can be thought of as its ability to separate two closely spaced ion beams. It can be defined in terms of geometrical parameters as:

$$\text{Resolution} = R = \frac{m}{\Delta m} = \frac{r}{S_1 + S_2 + \Sigma_a}$$

where S_1 and S_2 are the source and collector slit widths and Σ_a is the sum of the aberrations. It is interesting to note that Aston's machine had a resolution of about 130 which was adequate to discover many of the low mass isotopes, but a resolution of 10^5 is often required for mass measurements.

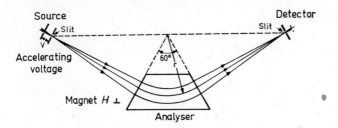

A sector magnetic analyser.

Figure 2 Simple mass spectrometer

2   BASIC INSTRUMENTS FOR NUCLEAR PHYSICS RESEARCH

Many types of mass spectrometer have been constructed since Aston's time, but this paper will briefly review some instruments of advanced design which are commonly used in nuclear physics applications.

The first of these instruments is the tandem magnetic type that was developed by Inghram and Hayden in 1954[5]. Two machines of this design, the

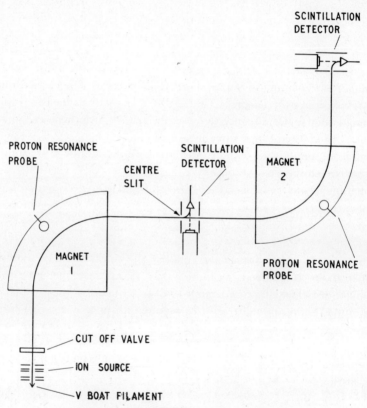

Figure 3   MSX two stage mass spectrometer

MSX and MSY, have been built at AWRE Aldermaston and the basic layout of one of them, the MSX spectrometer as described by Wilson *et al.* (1961)[6] is illustrated in Figure 3. These instruments are used primarily for the measurement of isotope ratios.

Material to be analysed is deposited on a V-shaped filament located in the ion source. The filament is heated, and some of the material is ionized by the surface ionization process. Ions formed are accelerated to 15 keV and enter a 15 in. radius 90° sector magnet and ions of preselected mass pass through a slit, a similar magnetic analyser, a final slit, and into an ion detector. Pressures of about 10^{-7} torr are maintained along the ion flight path. Two scintillation detectors as described by Daly in 1960[7], one at the centre and one at the end of the machine, measure the ion currents. They operate by deflecting ions onto a high voltage electrode and the secondary electrons which are produced are accelerated into a phosphor viewed by a photomultiplier. The centre detector is used when single stage operation is required. The magnetic fields in the two magnets are very nearly equal, so that ions of a given e/m pass through the machine. These magnetic fields are held constant by proton resonance control and ion beams of different e/m are selected by varying the ion source acceleration voltage.

Ion counting methods are used for beam intensity measurements. Printer scalers produce results that are fed to a computer which is programmed to include such things as electronic dead time corrections and produce a result with its standard error.

Complex machines such as these are built for two reasons: high isotopic abundance sensitivity and high absolute sensitivity in the sense that they can measure very small samples by means of counting techniques. Residual gas atoms along the flight path scatter the ions in a beam and the function of the second magnet is to separate again these gas scattered ions. Figures of abundance sensitivity such as 10^5 to 1 for single stage machines can, in principle, be elevated to 10^8 to 1 by the addition of a second magnetic analyser.

For the MSY machine the contributions in the uranium mass region from the main beam at adjacent mass positions on the low and high mass sides respectively are 1.3×10^{-6} and 3.6×10^{-8}. By the addition of a third electrostatic stage to this type of instrument White *et al.*[8] have shown that considerably higher figures can be obtained.

The sensitivity of these machines for handling small quantities of material can be illustrated as follows. Approximately 10^4 uranium atoms in the ion source are required to produce one detected ion. This means that if one wanted to measure a 1 to 1 ratio in uranium to an accuracy of approximately 1 %, only about 10^{-13} g would be needed. Some elements, such as caesium, are much more easily ionized than uranium with a consequent reduction in sample size.

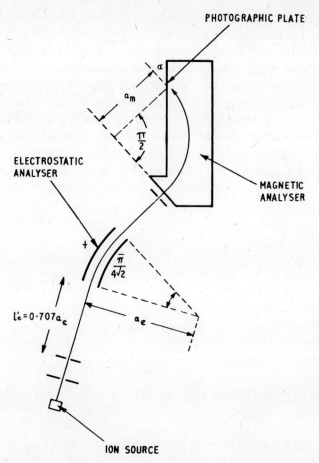

Figure 4   Geometry of the Mattauch–Herzog double focusing mass spectrometer

Double focusing mass spectrometers

The geometrical layout of one version of this design as developed by Mattauch and Herzog in 1934[9] is illustrated in Figure 4. Ions are formed in a high
voltage vacuum spark between pointed electrodes made of the material under
examination. This type of ion source is used since it gives approximately
equal ionization for all elements, but it also produces ions with a wide range
of energies. Ions are accelerated into an electrostatic analyser followed by a
magnetic analyser and different e/m values focus along a plane. Ion detection
is normally accomplished by placing a photographic plate along the plane

and detectable images are observed after the arrival of between 10^3 and 10^4 ions at any given mass position.

This ion source produces multiply charged ions and this property is useful in mass measurements. One can get mass doublets such as $^{52}Cr-^{12}C_2H_2$ or $^{54}Fe-^{12}C_2H_3$ when ^{52}Cr and ^{54}Fe are doubly ionized.

However, the main feature of this instrument is its ability to chemically analyse materials and also to detect minor constituents in that material down to 1 part in 10^9 under favourable circumstances.

Gas mass spectrometers

The third type of mass spectrometer is the gas analysis instrument in the form developed by Reynolds in 1956[10] at Berkeley for the measurement of extremely small samples.

Figure 5 shows the layout of an all metal Reynolds type machine. The whole machine, including the gas inlet system, is bakeable and getter ion

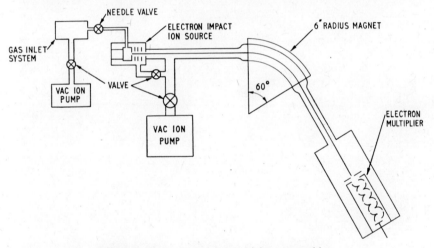

Figure 5 Layout of high sensitivity gas machine

pumps produce ultimate pressures of about 10^{-9} torr in the sampling system and 10^{-10} torr in the main analyser. The gas is leaked into the analyser through a needle valve and is ionized by electron impact. Since the ion current is measured with an electron multiplier the instrument has a high sensitivity for small samples. A resolution of approximately 500 is obtained with the 6 in. radius magnetic analyser. Stainless steel is used almost exclusively in this and the two previously described machines because of its good vacuum properties.

Ultra high vacua are needed to minimise the ion currents that arise from the residual gases in the system. Ion currents of about 10^{-16} A are recorded by this machine at the mass 18 (H_2O^+) peak. This is usually one of the highest residual gas peaks.

Instruments of this type can be run dynamically or statically. In the dynamic case the gas is admitted slowly into the ion source during the analysis.

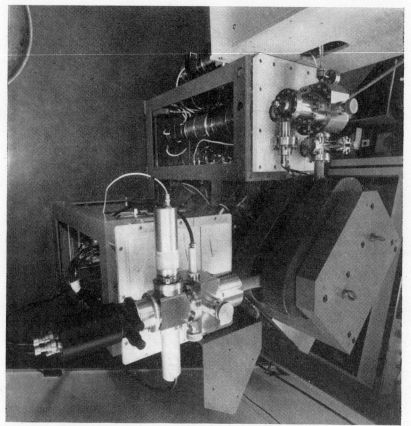

Figure 6   Photograph of the MSG high sensitivity gas mass spectrometer

In the static case the complete gas sample (if sufficiently small) is admitted, and it is then analysed. This second method produces lower ion currents, but they persist for considerably longer times than in the dynamic case. This method can be extremely sensitive, i.e. Reynolds showed that 5×10^5 atoms of a Xe isotope was detectable.

A version of the basic Reynolds type machine has recently been constructed at Aldermaston. The layout of this instrument is shown in Figure 5. It uses a 12″ radius 90° magnetic sector configuration, and is of all-metal construction, with gold gaskets throughout. The analyser section is pumped by an ion pump, but a fast double trapped oil pump is used on the ion source region. Bakeable metal ultra high vacuum valves are used, and the whole machine can be baked at 300°C. Ion counting methods are used with a scintillation detector, and a retardation lens system is built into the ion collector system. This device, described by Freeman *et al.*[11], is capable of enhancing the abundance sensitivity of single stage instruments by a factor of about 100. Collector and source slits are adjustable and a resolution of 3000 is easily achieved.

The basic requirement in the design of this instrument was to produce a machine that would be able to cope with a wide range of gas samples, from the ultra small to those requiring high resolution sufficient to resolve the C_2H_4–CO–N_2 triad.

Double collection mass spectrometers

An important instrument for the nuclear industry is one for measurements on ^{235}U and ^{238}U isotopes. Isotope separation plants have a requirement for high precision ratio measurements on uranium isotopes. The first instrument to make highly precise ratio measurements was developed by Nier *et al.*[12]. His introduction of double ion collector methods is illustrated in Figure 7, and the use of a balanced bridge method of comparing ion currents resulted in high measurement precision. The method has also been applied to oxygen isotopes and to the measurement of palaeotemperatures.

As part of the Nuclear Power Programme in Great Britain, highly precise mass spectrometers have been developed at Aldermaston to measure $^{235}U/$ ^{238}U ratios. Uranium is converted to UF_6 gas and a fairly conventional electron impact type of ion source is used. The electron beam region is made of copper, and is of open construction to reduce memory effects. Actual ratios are measured on UF_5^+ since this is the most abundant ion.

These machines are 12″ radius 90° sector instruments. They are designed to ultra high vacuum standards since they are constructed mainly of stainless steel and they are bakeable. They have run successfully with ion pumps

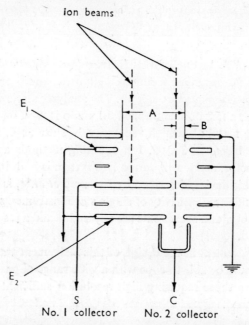

Figure 7   Schematic diagram of the double collection method

evacuating the ion source regions, and the ion flight path region. The electronic units are of modern design employing solid state components wherever possible.

Figure 8 shows a photograph of one of these machines. Natural uranium can be measured to an accuracy of 0.01 % on the ratio 235/238 which in this case is equal to 0.007.

Low intensity electromagnetic isotope separator

The design of most low intensity separators is based on that of the Scandinavian machines. The layout of a typical instrument used for muclear research at Aldermaston is shown in Figure 9.

The ions which are produced in an arc discharge ion source are accelerated through ~ 60 kV and mass analysis is achieved in a 100 cm radius, 90° sector magnet. Elements or suitable anhydrous compounds with a vapour pressure above 10 microns at a temperature below 1000 °C are heated to produce a vapour stream into the ion source. In some cases oxide samples are heated and converted to volatile chlorides by a stream of carbon tetrachloride vapour. Ion beams of the platinum group of metals can be formed by sputtering an electrode mounted in a gas discharge. The magnetic type arc discharge ion source has an efficiency of 1–20% and beams of 10–300 μA can be extracted from an aperture of about 2 mm diameter. The full acceleration volt-

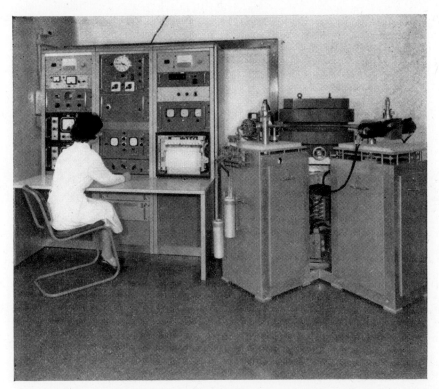

Figure 8 Photograph of a double collection mass spectrometer for precise measurement of U^{235}/U^{238} ratios using uranium hexafluoride

Table 1 Examples of enriched isotopes produced in the AWRE separator

Isotope	^{53}Cr	^{87}Sr	^{117}Sn	^{149}Sm	^{149}Sm	^{149}Sm	^{239}Pu
% Abundance in sample	9.5	7.02	7.6	13.84	13.84	13.84	95
% Abundance in separated isotope	99.9	99.994	99.9	99.56	99.975	99.99	99.997
Overall enrichment	9500	200,000	12,000	1400	25,000	89,000	1800
Amount collected (μg)	5	4	3	200	15	5	130

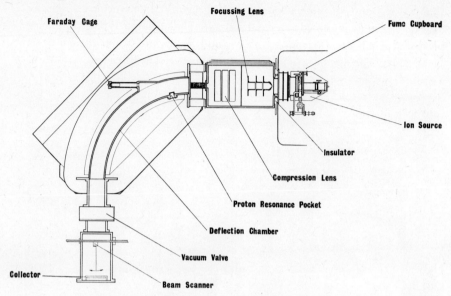

Figure 9   Low intensity electromagnetic isotope separator

age is applied across the gap between the ion source and the earthed extraction electrode, and the ions are focused into a slightly divergent beam by a spherical electrostatic lens. The height of the focused beams at the collector can be varied from approximately 1 to 50 mm by a cylindrical lens. The dispersion at the collector is R/M, where R is the radius of the machine, and M the mass number (i.e. 5 mm at mass 200) and a resolving power of 1000–3000 is usually achieved. The machine can produce separated isotopes in amounts up to 1 mg per day and enrichment factors of 1000 to 10,000 are readily attained (see Table 1).

The dimensions of the collector allow the simultaneous collection of isotopes in the mass range $M \pm 10\%$. Where possible the isotopes are collected in a form suitable for immediate use since this avoids possible contamination during chemical extraction. The direct collection of the ions at their full energy is the simplest method and is particularly useful for low beam currents. The amount that can be collected is limited by sputtering to a few $\mu g/cm^2$. The sample is not on the surface of the target foil since the ions penetrate to depths depending on the atomic numbers of the ion and target elements.

When it is essential to have a surface deposit the retardation technique of Sidenius and Skilbreid (1961) can be used[13]. If the ion energy is reduced so that the sputtering rate is low, deposits of at least 200 $\mu g/cm^2$ can be built up. The collector geometry can be arranged to eliminate the neutral particles produced by charge exchange, thus giving an improved enrichment.

Another technique is to allow the ion beam to strike a suitable target and collect the sputtered material (separated isotope + target material) on an adjacent foil. When large quantities are being collected the ions are focused into a deep copper pocket and the sample is purified by normal chemical methods or electrolysis.

3 APPLICATION IN NUCLEAR AND REACTOR PHYSICS

3.1 Nuclear physics

a *Half-life measurements of radioactive nuclides*

Half-lives can be determined in a number of ways using a mass spectrometer. One may examine the rate of disappearance of a parent nuclide or, alternatively, the rate of growth of a daughter nuclide.

In the first method it is necessary to have a stable long-lived isotope nearby to provide a reference for the measurements. The method can be used for half-lives that range from days to years although the accuracy will tend to fall off rapidly as the half-life increases. Thode and Graham[14] have studied the fission product gases. They measured the decay of ^{85}Kr relative to the stable ^{86}Kr. A half-life of 9.4 years was obtained and later remeasured as 10.27 years. Macnamara *et al.* in 1950[15] made a similar measurement on ^{133}Xe yielding a half-life of 5.27 days.

More recently (in 1964) Clarke and Thode[16] using a rapid irradiation of fissile materials and analysis of fission gases found for the half-lives of ^{87}Kr, ^{88}Kr, ^{135}Xe and ^{138}Xe values of 76.4 min, 2.80 hours, 9.19 hors, and 14.0 min respectively.

Dietz *et al.*[17] have made a study of the important nuclide ^{137}Cs using a two-stage mass spectrometer. Since it has a half-life of about 30 years measurements over several years are necessary to achieve high precision. To achieve this precision they use the "ratio of ratios" technique, whereby they measure the ratio of $(^{137}\text{Cs}/^{135}\text{Cs})(^{135}\text{Cs}/^{133}\text{Cs})$. This is a method of overcoming the fact that the transmission of ions of different mass varies with time. Measuring the decrease of ^{137}Cs with respect to stable ^{133}Cs through the intermediate long-lived ^{135}Cs results in a marked increase in the precision of the measurement. This important advance in the precision obtainable from mass spectrometers of the solid source type, has yielded a figure of 30.35 ± 0.38 years for the half-life of ^{137}Cs.

In the "daughter growth" method the amount of daughter formed in a known time gives the half-life. This technique has been applied to ^{235}U where the amount of ^{207}Pb present in old minerals has been measured. It has also been applied to ^{240}Pu giving a half-life of 6580 ± 40 years. Considerably longer half-lives can be examined using the daughter growth method than by the parent decay method.

A combination of counting techniques and mass spectrometry is an extremely effective method for the measurement of very long half-lives. The radiation decay equation is:

$$dN/dt = -N\lambda$$

where dN/dt is the disintegration rate, λ the decay constant, and N the number of atoms present. Counting methods give dN/dt and N can, in many cases be determined mass spectrometrically by isotopic dilution and hence the half-life is obtained. In 1961 McNair and Wilson[18] applied this technique to the measurement of the ^{87}Rb half-life.

In the isotopic dilution technique referred to above the mass spectrometer is effectively used as a very sensitive weighing machine. It can be applied to elements that have at least two stable or long-lived isotopes. Ideally, a known amount of single isotope is added to an unknown amount of a mixture of isotopes. By measuring the change this produces in the isotopic ratios one can determine the weight of the original sample. Quantities at least as small as 10^{-12} g have been measured in this way.

b   High energy nuclear reaction studies

The break up of an atomic nucleus under fast proton bombardment has been studied using mass spectrometric techniques. Knowledge of the formation cross section of fragments is useful for nuclear structure studies and theories of nuclear synthesis. Very sensitive solid source and gas machines have been used for this work. When solids are bombarded by fast protons and gas atoms are formed, they are extracted from the material by heating in a furnace.

Schaeffer and Zähringer's work in 1958[19] illustrates the type of information that can be obtained using these methods. Table 2 gives the relative amounts of some of the rare gases obtained by bombarding iron with protons of different energies.

Table 2 Spallation cross sections σ (mb)

Proton energy	^{3}He	^{4}He	^{21}Ne	^{22}Ne	^{36}A	^{37}A	^{38}A	^{39}A
0.16 GeV	11	120						
0.47 GeV	45	450	0.1		1	3.3	8	4.1
3 GeV	240	1300						
25 GeV	113	670	5.7	3.0	1.64	5.8	13.6	7.7

The last row of Table 2 refers to the results of Goebel and Zähringer[20].

Spallation is also produced by the interaction of cosmic rays with meteorites. The primary radiation consists mainly of fast protons and the energy distribution exhibits a maximum at about 1 GeV.

The number of atoms $n(A)$ of mass number A formed by spallation in Fe meteorites is given empirically by

$$n(A) = K_1(\Delta A)^{-k_2}$$

where ΔA is the mass difference between Fe and mass number A, K_1 is approximately proportional to the radiation dose at a point in the meteorite, and K_2 indicates the proton energy at this point.

Figure 10 compares the results obtained for the mass spectrum of potassium by Voshage and Hintenberger in 1961[21] for three iron meteorites with the mass spectrum of potassium obtained from proton bombarded iron and pure iron.

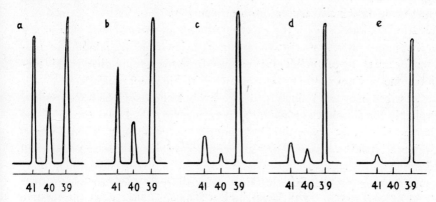

Figure 10 Mass spectra of potassium extracted from iron meteorites and from terrestrial iron, a, b, c are meteorite samples, d is pure iron after proton bombardment, e is pure iron

In 1963 Gradsztajn *et al.*[22] performed an interesting experiment where they examined the production of ^{6}Li and ^{7}Li formed in the bombardment of ^{16}O by 156 MeV protons. They irradiated an ice cube in the external beam of a synchrocyclotron and after irradiation the ^{6}Li/^{7}Li ratio was measured on a very sensitive mass spectrometer. Allowance was made for the small amounts of natural lithium in the water by making two irradiations of very different times. The absolute amount of Li formed was measured by isotopic dilution methods. It is interesting to note that the Li samples were as small as 10^{-12} g. They were analysed on a solid source machine equipped with an electron multiplier.

Results quoted by the authors are 9.8 ± 1.4 millibarns for the cross-section for ^{6}Li production, and $r = 1.4$ where:

$$r = \sigma\,(^7\text{Li})/\sigma\,(^6\text{Li})$$

A study of nuclides far removed from the nuclear stability line involves a study of short lived isotopes and produces information on delayed neutrons and protons, β- and γ-rays, α-particle emission, new double magic nuclei, deformed nuclei, Q values and relative reaction yields all of which contribute to our understanding of the nucleus. Generally speaking the half-lives of interest are much less than 100 secs. The study of neutron deficient or neutron rich isotopes produced by high energy particle reactions or by fission requires a rapid chemical and mass separation usually followed by sophisticated counting techniques in order to make unambiguous mass and atomic number assignments to the severals activities formed. A number of on line mass spectrometer and separator experiments are being used or are planned in various laboratories for the study of either high energy reaction products or fission products. A recent symposium[23] reviewed the techniques and reasons for the study of nuclides far removed from the stability line. In this section some on line experiments associated with high energy accelerators will be described in order to illustrate the variety of techniques which are applied. Probably the simplest approach has been made by the Orsay group under Bernas using fairly conventional 15 cm or 30 cm radius mass spectrometers.

Figure 11 shows a schematic representation of the on line experiment of Klapisch and Bernas[24]. These workers used diffusion in solids at high temperatures to extract reaction products from a target. Their ion source consisted of a succession of thin foils of the element under study and of thin graphite slabs enclosed in a heated cylindrical metal foil. A high energy proton beam crossed the ion source which was heated to between 1500° and

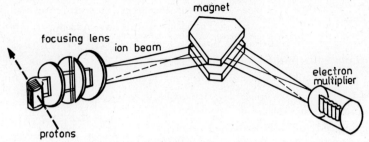

Figure 11 Schematic representation of the mass spectrometer (on line) operating
in the beam of an accelerator

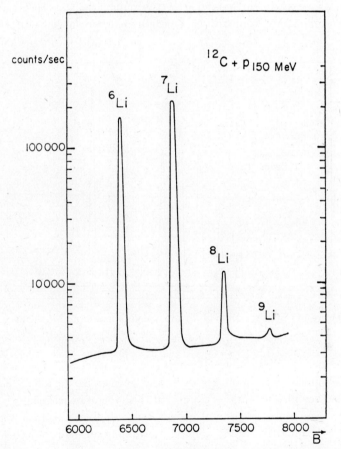

Figure 12 Li spectrum obtained from 150 MeV proton bombardment of a carbon
target ion source

1800°C. Reaction products recoil from the target, are slowed down in the graphite, and diffuse through it. The alkali metals have a high probability of leaving the graphite surface as positive ions.

The ion optics of the source was of conventional design, with an acceleration voltage of 3 kV and ion counting methods were used for recording the ion beams.

Figure 12 shows a lithium spectrum obtained from a simple ion source consisting only of carbon slabs. Here a 150 MeV proton beam was used to produce ^{6}Li and ^{7}Li, and, in addition, ^{8}Li and ^{9}Li with half-lifes of 0.8 and 0.17 secs respectively.

In another experiment Amarel *et al.*[25] examined the mass spectrum of rubidium obtained by bombarding a uranium-carbon target with 150 MeV protons.

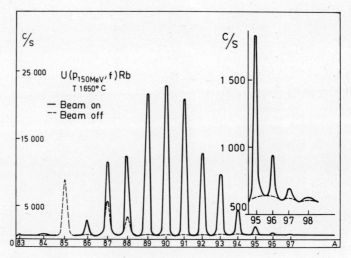

Figure 13   Mass spectrum of the rubidium isotopes formed in the fission of uranium by 150 MeV protons

This experiment has shown (Figure 13) the existence of three new isotopes of rubidium with mass numbers 96, 97 and 98, each having a half-life of less than one second. The authors claim that this type of experiment easily permits the accurate subtraction of background rubidium ions from those due to proton beam interactions, simply by switching off the proton beam.

The aim of this type of work is to produce cross sections for production of various isotopes and it can be seen that corrections have to be applied to al-

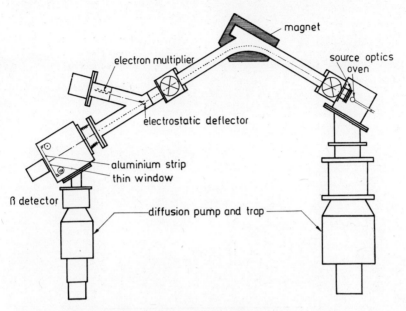

Figure 14   Diagram of the on line mass spectrometer used by Amarel *et al.* for studying the decay of short lived isotopes of Rb and Cs

low for such things as half-lives of short lived isotopes and for diffusion times. Cross section results obtained by these workers were in very good agreement with the results obtained by Friedlander *et al.*[26] using chemical separation followed by mass spectrometry. The sensitivity of the method is such that cross sections in the hundreds of microbarn region can be measured easily.

Another very interesting experiment carried out by the same workers was concerned with half-life measurements and the decay schemes of collected isotopes. Here Amarel *et al.*[27] bombarded 0.5 g of uranium with 150 MeV protons and the separated isotope beams were intense enough to study β- and γ-rays emitted from the radioactive nuclides.

Figure 14 shows the experimental arrangement, the main difference lying in the collector end region of the spectrometer. Electrostatic deflectors could deflect the ion beam to an electron multiplier, or to a thin metal strip behind which could be located α, β or γ counters.

Half-life measurements are made by switching on the proton beam for a predetermined time and then switching off both the proton beam and the

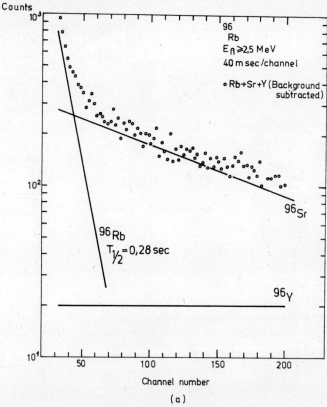

Figure 15   Typical decay curves for β-rays measured at the collector of the mass spectrometer in Figure 14

mass spectrometer ion beam and starting the β counter. A summary of the experimental results is given in Table 3.

A typical decay curve is shown in Figure 15, taken measuring β-rays from which a half-life was determined. Delayed neutron emitters were also detected using this experimental method.

In a recent experiment performed at the CERN proton synchrotron, Chaumont et al.[28] bombarded targets of Ta, Th and U with 24 GeV protons. The target was bombarded with 7×10^{11} protons every 11 seconds and the fast diffusing part of the reaction products, in the mass region of interest, was taken from the difference between two mass spectra, one recorded just after the proton burst and another 8 seconds later. A multiscaler was used to integrate the contributions of many bursts.

Table 3 Half-lives measured at the collector of the on line mass spectrometer of Amarel *et al.*; delayed neutron emitters (e) and β-emitters (m) not previously known

| | Half-life determination with on-line mass spectrometer | | |
Isotope	By neutron detection (sec)	By β-ray detection (sec)	Previous half-life measurement (sec)
^{92}Rb	none	4.40 ± 0.06	4.1 ± 0.3
^{93}Rb	5.81 ± 0.15	5.87 ± 0.06	5.1 ± 0.3
^{94}Rb	e 2.74 ± 0.06	2.63 ± 0.04	2.9 ± 0.3
^{95}Rb	e 0.36 ± 0.02	0.34 ± 0.02	≤2.5
^{96}Rb	e 0.21 ± 0.02	m 0.28 ± 0.05	none
^{97}Rb	e 0.135 ± 0.010	–	none
^{95}Sr	–	26 ± 1	33 ± 6
^{96}Sr	–	m 4.0 ± 0.2	none
^{142}Cs	e 3.3 ± 0.6	–	2.3 ± 2
^{143}Cs	e 1.69 ± 0.13	1.60 ± 0.14	2.0 ± 0.4
^{144}Cs	e 1.05 ± 0.14	m 1.06 ± 0.10	none
^{144}Ba	–	m 11.4 ± 2.5	none
^{144}La	–	m 41 ± 2	none

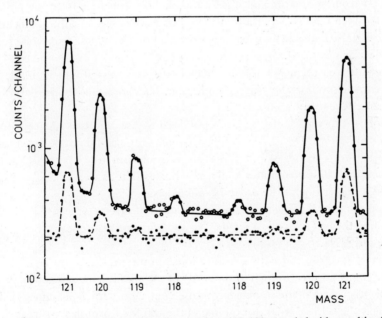

Figure 16 Mass spectrum of the mass region 118–121 recorded with a multiscaler by symmetrical triangular modulation of the accelerating potential. A first spectrum (open points) taken just after the proton burst (for some 100 msec) represents the fast diffusing part of the reaction products together with residual peaks which are measured separately as a second spectrum (solid points) 8 sec after the pulse

Table 4 Half-life measurements

Nuclide	This work half-life (sec)		Other measurements half-life (sec)
^{78}Rb		393 ± 11	360 ± 60
^{119}Cs		33 ± 8	
^{120}Cs		61.3 ± 1.4	
^{121}Cs		125.6 ± 1.4	120 ± 30
^{122}Cs	a)	21.0 ± 0.7	^{122}Cs: 0.50 ± 0.05
	b)	267 ± 11	
^{123}Cs		352 ± 3	336 ± 6
^{124}Cs	a)	26.5 ± 1.5	7200
	b) long-lived component		
^{126}Cs		98.6 ± 1.0	96 ± 12

Figure 16 shows the result obtained by these workers over the mass range 118–121 from bombardment of a Ta target. They conclude from the difference between the two spectra that the ^{118}Cs peak is statistically significant. ^{76}Rb was also detected as a spallation product of Ta and its formation cross section was measured.

Half-life measurements were made by letting the mass of interest impinge on the first dynode of the electron multiplier. The β-rays emitted are detected with an efficiency of about 20%.

Table 4 shows the half-life results obtained by use of a least squares computer programme. Decay curves for ^{122}Cs and ^{124}Cs could not be fitted to a single period. The authors interpret this as isomerism in ^{122}Cs or ^{122}Xe. In the case of ^{124}Cs a computer analysis indicates that as well as the 26.5 sec period a longer one of the order 12 min exists.

Klapisch *et al.*[29] using again the on line mass spectrometer technique have measured the half lives of ^{11}Li and ^{27}Na and of the new isotopes ^{28}Na, ^{29}Na, ^{30}Na and ^{31}Na.

A mass spectrometer with a surface ionization type of ion source as used by the Orsay group is limited to the study of the alkali element reaction products. Other types of ion source may be used but a more efficient and versatile system is the isotope separator on line to the 600 MeV proton synchrocyclotron at CERN[23].

It is obvious that the on line mass spectrometer is a very powerful tool in nuclear physics and it can be developed to study many more nuclides.

c   On line fission experiments

Mass spectrometers have been used for many years to study the fission process following the experiment of Nier *et al.*[30] in 1940, which showed that the [235]U isotope was fissioned by thermal neutrons. Many workers have examined fission product yields by radioactive counting methods and by isotope

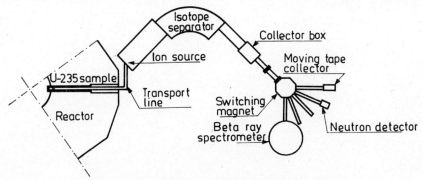

Figure 17   Layout of the on line separation system at the
Ames National Laboratories

dilution mass spectrometry. By now the yield curve is mapped out in considerable detail and attention has been turning to examining short lived fission products. To do this a series of on line experiments have been built up. Here the products of nuclear fission are quickly swept into an isotope separator where mass separation is achieved and the separated atoms are examined by counting methods to determine their decay schemes.

A good example of one type of experiment is illustrated by the work at Iowa State University by Talbert *et al.*[23,31]. Figure 17 shows the experimental layout. A powdered oxide sample of [235]U is located in an evacuated tube near the core of a reactor. The tube is 5 metres long and is brought out through shielding and then enters the ion source of an isotope separator. Measurement indicates that the flow time of gas from the sample to the ion source is a few seconds under the molecular flow conditions that prevail. The whole tube can be heated to assist the transport of less volatile fission products. After passing through the mass separator the beam can be deflected by a switching magnet into any of five experimental chambers containing nuclear spectroscopic equipment.

The method has been applied to study delayed neutron emissions from the daughters of selected noble gas isotopes.

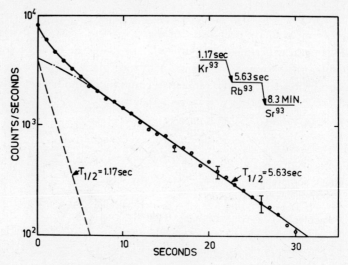

Figure 18   Time dependence of the counting rate of delayed neutrons from mass 93 isotopes

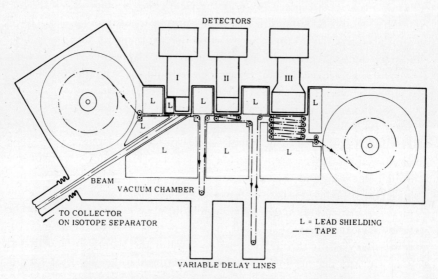

Figure 19   Schematic of detector box with moving tape assembly

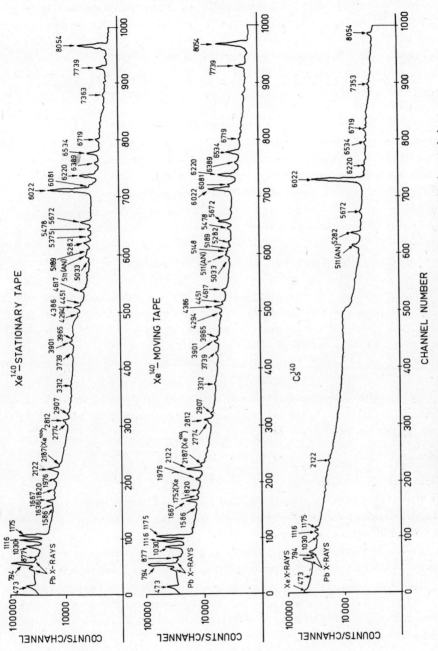

Figure 20 γ-radiation up to 2 MeV observed from the mass 140 decay chain

Figure 18 shows the time dependence of the counting rate of delayed neutrons from isotopes collected at mass 93 where ^{93}Kr and ^{93}Rb are shown to be precursors of delayed neutrons.

In another somewhat similar experiment carried out by the Princeton group[32] the radiations from ^{140}Xe ($T_{1/2}$ = 16 sec) and ^{140}Cs ($T_{1/2}$ = 66 sec) have been examined.

A source of ^{252}Cf was located near the ion source of the separator and it had a fission rate of 1.5×10^8 fissions/minute. The fission products from the sample were stopped in barium stearate. These rapidly leave the stearate and are swept by a stream of xenon into the ion source. The mass analysed ion beams were collected on a moving tape arrangement shown in Figure 19. Germanium counters monitored the tape at various times and the tape speed could be varied. By this means various γ-ray lines could be assigned to the decays of different members of the selected mass chain and half-lives determined.

Figure 20 shows a part of the γ-ray spectrum connected with the decay of ^{140}Xe ($T_{1/2}$ = 14.3 sec) and its daughter ^{140}Cs (65.7 sec). In the top spectrum the tape was stationary and for the middle it was moving. The intensities of the peaks in the ^{140}Cs decay are relatively diminished, particularly the 602.2 keV line. The lowest spectrum corresponds to position II and shows the lines following the decay of ^{140}Cs.

d   Atomic mass and weight determinations

Very accurate mass measurements are needed by the physicist to work out binding energies and by the chemist to calculate atomic weights. The binding energy per nucleon, B, can be defined as:

$$B = \frac{(ZM_H + NM_N - M^1)\,c^2}{A}$$

where c is the velocity of light, M_H and M_N are the masses of the hydrogen atom and neutron, and M^1 represents the mass of an atom of mass number A and atomic number Z. Here $N = A - Z$.

Figure 21 shows the binding energy as a function of mass number up to mass 240. Structure is very apparent at the low mass end with a periodicity of four mass units. This demonstrates the applicability of the Pauli exclusion principle in nuclei because each four-shell contains just two neutrons and two protons. A break at mass A = 90 is attributed to the completion of the 50 neutrons shell.

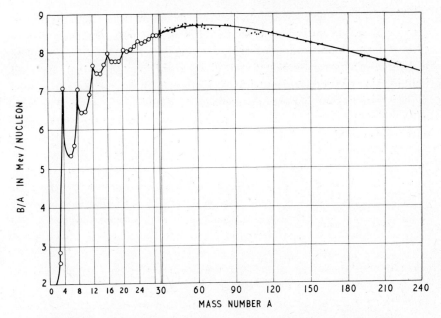

Figure 21   Binding energy in MeV/nucleon as a function of mass number

Atomic masses can also be obtained using nuclear reactions and measuring Q-values where Q is defined as the difference between the ground state masses of the interacting nuclei before and after the reaction. In other words:

$$Q = M + N - M^1 - N^1$$

If Q can be measured this method can be combined with mass spectrometry measurements to yield unknown masses very accurately.

The chemist is interested in an accurate measurement of nuclear mass and isotopic composition in order to work out atomic weights. A simple example of an atomic weight measurement is illustrated below.

	Atom (%)	Nuclidic mass
^{85}Rb	72.15	84.911710 ± 60
^{87}Rb	27.85	86.909180 ± 80

$\therefore$ Atomic weight = $84.9117 \times 0.7215 + 86.9092 \times 0.2785 = 85.4680$

The geometrical configuration of a typical mass measuring machine is illustrated by Nier's[33] instrument shown in Figure 22. The machine is of the double focusing type and employs electrical detection rather than the photographic plate. The mass doublet technique is used to measure masses. Examples of mass doublets are $^1H_2-^2D$, $^{12}C^1H_4-^{16}O$,$(^{16}O)^+-(^{48}Ti)^{+++}$. The technique of comparing heavy multiply charged ions with singly charged light ions is a very convenient trick used in this field. A peak matching technique is used to superimpose one member of a doublet on another. The voltage required to do this is measured and is proportional to the mass difference in the doublet.

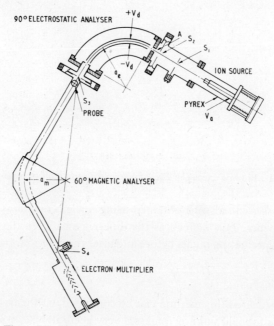

Figure 22   The Nier double focusing mass spectrometer

Figure 23 shows the complexity of peaks one can obtain at mass 16. This beautiful plate was obtained by Mattauch with a machine having a resolution of 100,000. Present day techniques can measure mass with a precision of at least 1 part in 10^7.

e   Branching ratios

Mass spectrometry can be applied to the measurement of branching ratios if isotopic dilution methods are used. Consider the case of a nuclide that can proced $Z \rightarrow Z + 1$ (β emission) or $Z \rightarrow Z - 1$ (β^+ emission or K-electron

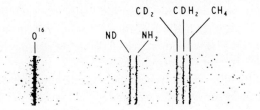

Figure 23 Multiplet at mass number 16 (by courtesy of Prof. J. Mattauch)

capture). Then the branching ratio λ is given by

$$\lambda = \frac{N(\beta^-)}{N(\beta^+ + E^c)}$$

This method was applied by Inghram *et al.* in 1950[34] to determine for the important nuclide ^{40}K. The amount of ^{40}Ca and ^{40}A of radiogenic origin in an old potassium mineral was measured using ^{36}A and ^{48}Ca as the dilution isotopes.

Separation of the effects of β^+ decay and electron capture can be achieved if a simultaneous measurement is available by counting methods of $N(\beta^+)/N(\beta^-)$.

f Fission yields

The general shape of the slow neutron induced fission yield curves with their broad mass range centred at $A = 95$ and $A = 140$ has been known for a considerable time. The first yields were measured by radiochemical methods but since then much work has been accomplished using mass spectrometers.

Fission fragments are formed in neutron enriched states and decay towards the Bohr–Wheeler stability line by a series of β-transitions. These activities will eventually decay to stable isotopes whose abundance can be measured to obtain relative fission yields. Most emphasis in this field has been placed on ^{235}U yields but data are also available for ^{232}Th, ^{233}U, and ^{239}Pu. Forty-three mass chains have been examined and these, with some chains determined radiochemically, have led to the absolute yields of the various mass chains to an accuracy of $\pm 3\%$. In 1958 Katcoff[35] tabulated the low energy neutron induced fission yields for ^{235}U, ^{232}Th and ^{239}Pu.

One of the earlier results that came from a study of fission yield was the discovery of a krypton isotope at mass 85. Figure 24 shows the mass spectrum of fission product krypton as obtained by Thode and Graham in 1947[36].

Before this mass spectrum of a 1–2 year old fission gas sample was taken, the only known ^{85}Kr isomer had a half-life of 4 hours. By measuring the ^{85}Kr isotope at intervals of several months its half-life was found to be 10.27 years. The yield of this isomer was also obtained.

One of the difficulties in this type of measurement is that since all isotopes capture neutrons to some degree, the mass yield curves can be distorted by neutron capture.

In 1955 Wanless and Thode[37] minimized this effect by irradiating at low fluxes and measured the relative abundances of krypton and xenon isotopes. Table 5 shows the results obtained.

The mass yield curve for thermal neutron fission of ^{235}U is shown in Figure 25. The curve has been folded over so that complementary masses coincide assuming that 2.5 neutrons are emitted per fission. It is possible knowing the mass yield curve to calculate the average number of neutrons per fission $(\bar{\nu})$ from:

$$\bar{\nu} = 236 - 2\left(\frac{\Sigma \, \text{mass} \times \text{yield}}{\Sigma \, \text{yield}}\right)$$

A value 2.8 ± 0.08 for ^{239}Pu obtained by this method is in very close agreement with the accepted value of 2.90.

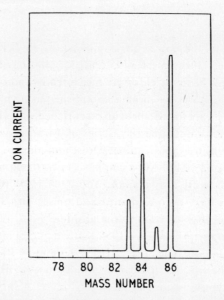

Figure 24   Mass spectrum of fission product krypton

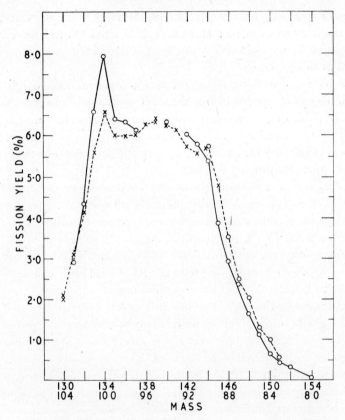

Figure 25   Mass yield curve for thermal neutron fission of ^{235}U

Table 5   Fission yields for Kr and Xe isotopes

Isotope	Atom abundance (%)	Absolute fission yields (%)
^{83}Kr	14.10	0.577
^{84}Kr	25.93	1.02
^{85}Kr 10.27 hr.	7.59	0.300
^{86}Kr	52.38	2.07
^{131}Xe	13.42	2.93
^{132}Xe	20.08	4.38
^{134}Xe	36.91	8.06
^{136}Xe	29.59	6.46
^{133}Xe 5.27 days	^{131}Xe/^{133}Xe (0.443)	6.59

Mass spectrometry has also revealed the structure in the mass yield curve. An example of this can be seen as a peak at mass 134 in Figure 25. It is believed that the special stability of the 82 neutron shell causes the build-up of yield at this point.

Since the most probable fission product in a given mass chain is located early in the chain where half-lives are short not much information has been obtained concerning the primary fission products. However, in special cases where shielded nuclei exist as in the cases of ^{90}Br and ^{128}I the yields have been measured. They have been found to be 10^3–10^4 times less than those of the most probable primary products.

In a very recent experiment Laeter and Thode[38] have examined the relative yields of tin isotope in neutron induced fission in ^{233}U. They used a high sensitivity tandem mass spectrometer and electron impact ionization techniques to analyse 10^{-9} g samples of tin.

Few measurements have been made in the symmetric region of the fission yield curve, mainly because the yields are 100 times lower than at the peaks for thermal fission, although the situation improves for fast fission. They were also interested to see if fine structure effects could be detected in this region. Tin with its ten stable isotopes, many of which are the end of fission chains seemed ideal since it covered a large part of the symmetric mass region.

Samples were irradiated in the form of U_3O_8 in evacuated quartz capsules. Nanogram quantities of fission produced tin were extracted from the irradiated uranium sample using anion exchange techniques. The sample was then put in an electron impact crucible type source as described by Tyrell et al.[39] and the sample analysed on the tandem mass spectrometer using ion counting techniques. The spectrum showed tin isotopes and hydrocarbons, and the resolution was sufficiently good to separate these.

Table 6   Relative yields of tin isotopes

Isotope	^{233}U (thermal fission)	^{233}U (fast fission)	Fuel rod (^{235}U + ^{238}U + ^{239}Pu)
^{117}Sn	0.180 ± 0.013	0.21 ± 0.02	0.204 ± 0.006
^{118}Sn	0.186 ± 0.013	0.21 ± 0.02	0.196 ± 0.006
^{119}Sn	0.189 ± 0.014	0.26 ± 0.03	0.213 ± 0.006
^{120}Sn	0.210 ± 0.014	0.29 ± 0.03	0.228 ± 0.006
^{122}Sn	0.234 ± 0.013	0.29 ± 0.03	0.254 ± 0.005
^{124}Sn	0.386 ± 0.020	0.42 ± 0.04	0.427 ± 0.009
^{126}Sn	= 1.000	= 1.00	= 1.000

Table 6 shows the relative yields of tin isotopes obtained and these values have been corrected for mass discrimination effects caused mainly by the electrostatic scanning techniques used. Corrections amounting to 1.5% per mass unit were applied, and this value was obtained from the use of standards. Uncertainties, these workers suggest, come mainly from the terrestrial tin contamination. The figures in the second column for fast fission were obtained by wrapping some samples in cadmium which, of course, strongly absorbs the thermal neutrons.

Figure 26 shows the mass-yield curve obtained, and despite the prediction by Third and Tomlinson[40] that fine structure should be present, it seems to be absent. The yields are essentially constant from 117 to 122, and then increase fairly rapidly.

Meek and Rider[41] have recently summarised fission product yield data.

In another recent experiment Srinivasan *et al.*[42] have measured Xe and Kr isotopes from the spontaneous fission of californium 252. Here two aluminium foils 0.001 inch thick by 0.75 inch diameter were exposed to a califor-

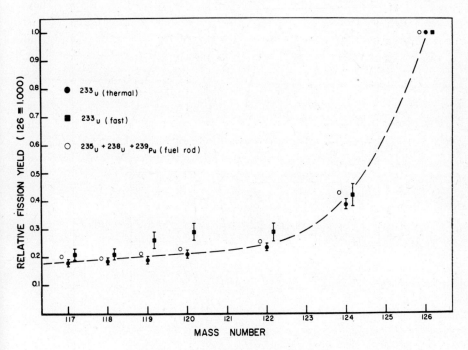

Figure 26 Relative fission yield of tin isotopes

nium source ^{252}Cf of approximately 1×10^7 fissions per minute. After 5 months and an integrated flux of 5×10^{10} fissions, they were mounted in a glass bottle and the pressure reduced to 10^{-9} torr. The foils were then dropped into a previously outgassed molybdenum crucible and melted by induction heating. After extracting the Kr and Xe, the gases were measured on a Reynolds type mass spectrometer. Various corrections were applied to the results obtained to allow for mass discrimination and atmospheric contamination. The atmospheric component was calculated by assuming that all the ^{82}Kr and ^{129}Xe were of atmospheric origin since they are shielded from β-decay by stable ^{82}Se and 17 million year ^{129}I.

Figure 27 shows the results obtained and a comparison with the results of Schmitt[43], Newick[44] and Glendenin and Steinberg[45]. There is evidence of fine structure at mass 132 and mass 134, as is also observed in neutron induced fission of ^{235}U.

Yields such as these are of geological importance because of the possible existence of such elements in the early history of the solar system.

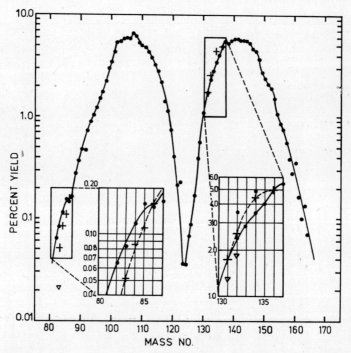

Figure 27　Yield of Kr and Xe isotopes from the spontaneous fission of ^{252}Cf

Much work yet remains to be done on such things as mass yield variation with neutron energy and on the yield of the primary fission products themselves.

g   Nuclear studies with separated isotopes

The use of enriched stable isotopes as targets in nuclear reactions reduces the number of possible reactions and, besides simplifying the analysis of the experimental data, the masses of the reaction products can usually be determined unambiguously. Measurements with neutrons and high energy particles often require amounts in excess of 100 μg and these are supplied by the Calutrons at ORNL and Harwell. The quantities produced in a low intensity separator are generally sufficient for experiments with low energy accelerators, e.g. the Van de Graaff. Metallic targets of high isotopic purity with a thickness of 50 μg/cm^2 can be collected on 50 μg/cm^2 carbon films by the retardation technique. The advantages of these thin high purity targets for Coulomb scattering experiments are illustrated by the experiments of Eccleshall *et al.*[46] who studied the Coulomb excitation of ^{130}Ba, ^{52}Cr, ^{54}Fe and some of the even mass isotopes of Sm, Nd, and Ce by measuring inelastically scattered oxygen or sulphur ions in a magnetic spectrometer, or by recording coincidences between back scattered ions and γ-rays from the Coulomb excited levels.

Monoisotopic gas targets can be prepared by allowing energetic ions to strike a collector foil. At 50 keV the gas atoms are collected at depths of up to 20 μg/cm^2 in low-Z materials and gas targets with a thickness equivalent to a few μg/cm^2 can be made. Brostrøm *et al.* in 1947[47] were the first to use this technique in the study of Ne (p, r) Na reactions and, in 1952 with similar targets, Mileikowsky and Whaling[48] measured the Q of the reaction ^{21}Ne (d, p) ^{22}Ne.

Optical spectroscopy, atomic beam resonance, and radiofrequency methods are used to determine important nuclear parameters such as spin, magnetic moment, and electric quadrupole moment. It is usually impossible to make accurate measurements unless highly enriched isotopes are available.

h   Sources for α, β, and γ-spectroscopy

Radioactive sources produced in an electromagnetic separator have many advantages. If an adequate chemical separation is made prior to mass separation an unambiguous assignment can be made to the nuclide. The collection of ions on low-Z foils produces thin high specific activity sources with low self absorption and scattering of the emitted radiations. If the penetration of the ions is sufficient to distort the spectra of α-particles or low energy β-rays and conversion electrons (Figure 28)[49] the sources can be prepared by the retardation technique.[13] Even when the mass number of the active isotope is known it is advantageous to make a carrier free source. For example, high resolution spectroscopy is impossible with the low specific activities produced in (n, γ) reactions, particularly when the cross section is low.

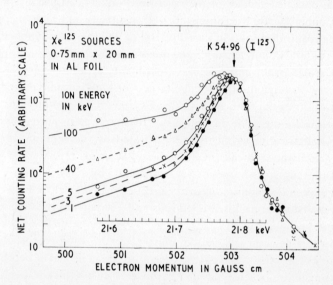

Figure 28   Distortion of the K 54.96 keV conversion line from Al targets implanted by ^{125}Xe ions at various energies

The high separation factor of the research separator is essential where mixtures of isotopes with large variations in activity are encountered; again, the high efficiency of these machines reduces the amount of activity to be handled and simplifies the contamination problems. The speed of separation can be quite short and a few minutes is often sufficient to give adequate source strengths. For the shortest half-lives counters or scintillation spectrometers have been mounted directly on the collector (e.g. on line separators).

i   The mass assignment of radioactive nuclides

The identification of the masses of natural radioactive isotopes was accomplished by mass separation techniques. In 1937 Smythe and Hemmendinger[50]

used enriched material to identify ^{40}K and ^{87}Rb as the active isotopes of naturally occurring potassium and rubidium respectively.

The problem is more complex when fission isotopes or the products of high energy reactions are investigated. A chemical and isotopic separation followed by activity measurements is a powerful technique for studying these reactions. The separator must have a high resolution and dispersion to give adequate purity. Collection over a wide mass range and a high efficiency are desirable since the number of active atoms may be quite small. The points outlined under (h) apply in this type of work.

The identification and half-lives of many rare earth isotopes produced in the fission of ^{235}U was accomplished in Chicago in 1946–1948 using a Nier-type 60° mass spectrograph[51] with a surface ionization source. A similar apparatus was used during 1949–1950 at the University of California to study neutron deficient isotopes produced by high energy α-particle reactions. The first use of a separator to study spallation reactions was by Anderson in 1954[52]. Vanadium bombarded by 187 MeV protons was found to contain 12 nuclides of Ar, K, Sc, Ti, and V with half-lives ranging from 20 minutes to 3 days.

In some cases the range of elements produced in a reaction allows a partial separation to be achieved by fractional evaporation of the target in the separator furnace. Uhler *et al.*[58] used this technique recently when studying the α-activities of neutron deficient isotopes produced by bombarding Tl, Pb and Bi with 100 MeV carbon ions.

Rudstam[54] has used the CERN separator to measure the primary fission yields of iodine isotopes in the fission of uranium by 600 MeV and 19 GeV protons. The efficiency of the chemical and isotopic separation was obtained by an isotopic dilution measurement with an ^{131}I spike. Absolute yields were determined by comparing the reaction ^{27}Al $(p, 3pn)$ ^{24}Na of known cross section with the yield for the shielded iodine isotope ^{130}I.

3.2 Reactor physics measurements

a Neutron capture and fission cross section measurements

When an atom captures a neutron to form a stable or long lived nuclide it is possible to measure the capture product using a mass spectrometer. If the total neutron flux that the parent isotope has seen during the neutron irradiation is known, the capture cross section for that nuclide can be measured. This picture is over simplified as in many instances the capture product strongly absorbs neutrons, but in cases where burn up of the product is small, it can yield accurate results.

Very often the technique is combined with the use of highly enriched isotopes produced by mass separators, this being necessary when an abundant isotope already exists at the capture product mass number.

The effect was first shown by Dempster in 1947[55] for the important neutron absorbing element cadmium. His analysis revealed that the isotope ^{113}Cd was the predominant neutron absorber in the cadmium isotopes and that the decrease in ^{113}Cd was balanced by the rise in ^{114}Cd.

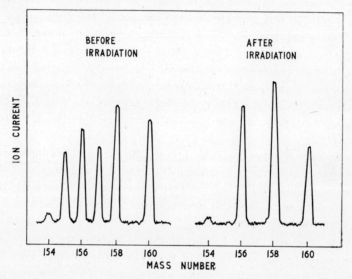

Figure 29   Mass spectrum of Gd before and after thermal neutron irradiation

Gadolinium is an element that has two isotopes, ^{155}Gd and ^{157}Gd, with neutron capture cross-sections of 56.200 and 242,000 barns respectively.

Figure 29 shows the mass spectrum of gadolinium found by Walker in 1954[56] before and after irradiation. It can be seen that ^{155}Gd and ^{157}Gd have been transformed to ^{156}Gd and ^{158}Gd. Neutron flux measurements that are necessary can be carried out by some of the normal methods such as activation of ^{59}Co.

An interesting combined mass spectrometer and isotope separator was designed by White *et al.* in 1961[57]. Using this instrument they measured simultaneously the neutron cross sections of isotopes of two different elements. A tungsten filament was put at the collector end of a large double focusing mass spectrometer and predetermined very pure isotopes of several elements were shot into the filament with kilovolt energies. The filament was then irradiated and put into the ion source of the machine and the capture isotopes measured.

More recently Forman and White[58] have used a four stage mass spectrometer to measure the thermal neutron capture cross section of ^{147}Sm. Highly pure samples, as small as 10^{-10} gms were irradiated for the measurement and the integrated neutron flux and neutron temperature were monitored by the burn up of ^{149}Sm and ^{157}Gd.

Promethium-147 (β-active with $T_{1/2} = 2.67$ year) is a significant reactor poison, particularly in modern reactors in which the fuel is taken to high burn up values. The fission yield from ^{235}U is 2.5% and neutron capture leads to 5.4 day ^{148}Pm and 41.5 day ^{148}Pm, the thermal neutron cross sections being about 100 barns for each reaction. The latter isomer has a high cross section ($\sim 29,000$ barns) and other products of this capture decay series have high cross sections. Fenner and Large[59] have used mass spectrometry techniques to measure the neutron capture cross sections. The stable samarium daughter products formed during and after irradiation in a reactor amounted to about 1 µg and were chemically separated from the highly active ^{147}Pm (5 curie). The thermal and resonance integral cross sections were derived from the relative abundance of the samarium isotopes measured in a two stage mass spectrometer (assuming the known half-life of ^{147}Pm). The measured thermal cross sections were 72 barns for ^{147}Pm (n, γ) ^{148}Pm, 82 barns for ^{147}Pm (n, γ) ^{148}Pm and 22,000 barns for ^{148}Pm (n, γ) ^{149}Pm.

Apart from the measurement of neutron capture cross sections it is also possible to measure $\alpha = \sigma_c/\sigma_f$, i.e. the ratio of capture to fission for fissionable nuclei such as ^{235}U, ^{233}U, ^{241}Pu, etc. This can be achieved by a combination of mass spectrometry and radiation counting where σ_c and σ_f are measured by these methods respectively. This type of study is being made at Aldermaston.

Recently measurements have been made on very pure ^{239}Pu samples that have been irradiated in fast reactor fluxes and it is interesting to see the range of nuclear data that can be obtained quickly from one sample. Besides measuring α by mass spectrometry and radiation counting, it has been possible to measure the σ $(n, 2n)$ for ^{239}Pu. About 10^{-5} of the ^{239}Pu undergoes this reaction, and the measurement is made using a high abundance sensitivity tandem instrument. It is also possible to look at successive capture in the

^{239}Pu sample

$$^{239}\text{Pu} + n \rightarrow {}^{240}\text{Pu} + n \rightarrow {}^{241}\text{Pu}$$

From the ratio of ^{239}Pu to ^{240}Pu to ^{241}Pu it is easy to measure the relative capture cross sections and, if the flux is known, to get the cross sections absolutely.

Burn up in reactors can be determined by measuring the amount of some given fission product formed.

Work is also in progress to make a pure mass spectrometric measurement of α for some isotopes. In this method the total burn up of an isotope is measured to yield $(\sigma_c + \sigma_f)$ and the increase of the capture product yields σ_c and therefore α can be calculated. Much greater precision is needed for this method than can be obtained by normal techniques. However, the combination of ion counting and the ratio of ratios method gives a precision of better than 0.1 % and this should give an accuracy of about 1 % to 2 % in the value of α. Two stage mass spectrometers are especially suited to this type of measurement.

Capture cross sections for gases can be measured using a mass separator and a mass spectrometer. Ions of the required mass number are injected into thin foils at energies around 50 keV. Since the foils will only accept quantities of gas ions amounting to microgrammes/cm^2 before they begin to sputter off as quickly as they arrive, very sensitive gas mass spectrometers must be used. The foils are irradiated and the gas is baked out of them into a gas mass spectrometer. The measured isotope ratios and neutron flux give σ_c.

b   Measurement of (n, α) cross sections

Gas mass spectrometry has found applications to material problems that arise in high temperature thermal and fast reactors. Steels of various types are used for fuel element cladding and as part of the structure of the reactor. Neutron interactions of the (n, α) type produce helium in the metal and affect the mechanical properties such as ductility of the steels. In thermal reactors the main effect comes from the high capture cross section for thermal neutrons by boron; in fast reactors it arises from threshold type (n, α) reactions involving the major constituents.

Farrar[60] has used a spark source mass spectrometer to examine He produced in various metals at very low levels. However, he found that even by using standards the accuracy of a single determination was only 20–30 % and that serious contamination of the mass 4 position occurred from multiply charged ions of C^{3+} and O^{4+}.

Recently Freeman *et al.*[61] have been measuring a cross section σ_F averaged over a fission neutron spectrum for a number of elements, where

$$\bar{\sigma}_F = \frac{\int \phi_F(E)\,\sigma(E)\,dE}{\int \phi_F(E)\,dE}$$

Highly pure samples were irradiated in the Dounreay Fast Reactor. The purity was checked by spark source mass spectrometry and the samples were mounted in a niobium carrier forming part of a removable fuel element.

Figure 30 shows the experimental arrangement used for the helium analysis. A vacuum fusion furnace at a temperature of about 1600 °C can be loaded with six samples and these can, in turn, be dropped into the furnace by means of a small external magnet. A Nuclide Analysis Associates U.H.V. mass spectrometer is used to measure the helium released.

Figure 31 shows the helium released against time for a sample. For total fluxes of the order of 10^{21} neutrons/cm^2 helium amounting to 10^{-6} cc at NTP is produced.

Freeman *et al.*[61] have measured the (n, α) cross sections (averaged over a fission neutron spectrum) for nickel, iron, chromium and molybdenum using this method. The values are in the millibarn region. Once the cross sections have been established it is possible to predict He production in a variety of steels and alloys used in reactor construction.

c   *Impurity analysis*

In the construction of nuclear reactors it is often necessary to know very exactly the constitution of materials that will be irradiated by large neutron fluxes. This information is required for two reasons. Some elements such as gadolinium with a neutron absorption cross section of 44,000 barns would, if present as a small impurity in, for example, reactor graphite, result in a large neutron loss. Other impurity elements such as cobalt have to be kept at a low level because of their health hazard.

Highly sensitive impurity analysis is also necessary for the semiconductor materials such as silicon that are extensively used in solid state counters.

The analysis of materials for small contaminant levels can be performed by three mass spectrometric methods. One can often use isotopic dilution methods, but this involves chemistry and for monoisotopic elements it cannot be used.

Another method involves the use of a double focusing spark source mass spectrometer which can rapidly yield an almost complete picture of the con-

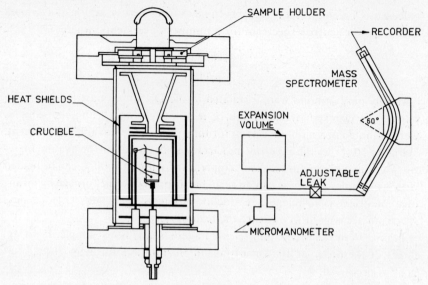

Figure 30   Schematic of vacuum fusion furnace and mass spectrometer for helium analysis

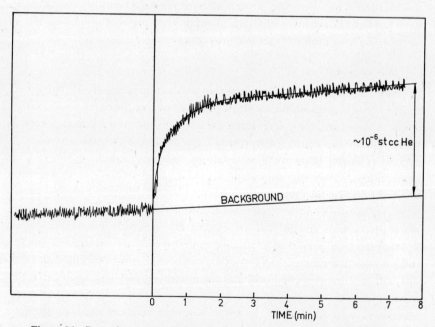

Figure 31   Recorder trace of helium released from a 10 mg chromium sample

stituents of a material very often down to levels of 1 part in 10^9. This machine has been briefly described already. Its application to the analysis of reactor graphite is illustrated by the results shown in Table 7.

Table 7 Analysis of a reactor graphite sample

	Sample ppm atomic	Neutron capture cross-section in barns
Cadmium	0.001	2400
Boron	0.3	750
Silver	0.002	60
Titanium	0.5	5.6
Vanadium	0.05	5.1
Nickel	0.02	4.5
Chromium	0.03	2.9
Iron	2.0	2.4
Calcium	3.0	0.43
Silicon	5.0	0.13

An AEI MS7 spark source machine was used for these measurements. It can be seen that the high capture cross section for the cadmium impurity is offset by its low (1 part in 10^9) concentration.

A lot of interest centres at the moment on stainless steel since it is used as a fuel canning material for the higher temperature reactors.

A mass spectrum of stainless iron, again produced by an MS 7 machine, is shown in Figure 32. The plate is made up of graded exposures covering a range 0.001 to 100×10^{-9} C in fifteen exposures. Quantitative measurements of the ion densities on the plate are made with a microdensitometer. As well as lines corresponding to other elements such as Sn and Pb, the complexity of the spectrum is increased by lines due to multiply ionized atoms such as Fe^{++} and Sn^{++}. This type of interference can reduce the sensitivity for detection of impurities.

The field of application of spark source double focusing mass spectrometers is rapidly expanding and doubtless will be extended in the sphere of nuclear technology.

Another interesting technique that is showing considerable promise as an analytical method for solids analysis is the sputter ion source. Various workers including Hönig[62] and McHugh and Sheffield[63] have studied ion pro-

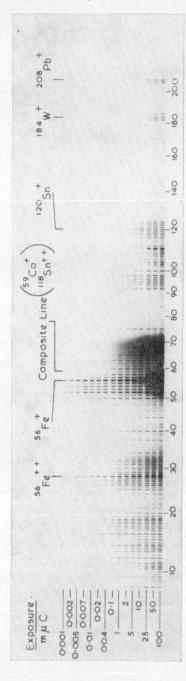

Figure 32 Mass spectrum of stainless iron (by courtesy of GEC/AEI Ltd., Manchester)

duction from surfaces that are bombarded by energetic ions. Surfaces are bombarded in good vacuum conditions with high density mass resolved ion beams. It is essential if contamination is to be kept to a low level that the surface is sputtered away at a much faster rate than background gases can adsorb on the surface.

The high current density is also required to produce adequate ion beams for mass analysis. As with most ionization techniques the efficiency is variable from element to element, and also matrix dependent. Herzog *et al.*[64] claim that the ion yield varied between 1 and 0.001 per incident ion and the neutral yield varies between 10 and 0.1. With primary ion beams of 100 μA this yields quite high beam intensities. Various ions have been used as the primary ion beam. Recent results by Andersen[65] suggest that one of the difficulties of using inert gas ions to bombard the surface, (i.e. the fall off of surface ion beam intensity with time) can be overcome if a reactive ion such as O^+ is used. It should be emphasised that the whole phenomenon of ion production from surfaces is a very complicated one and is affected by many variables such as residual gases present in the system, surface chemistry, surface structure, the matrix in which the element under investigation finds itself etc. Because of these effects it is essential to use standards to calibrate the instrument if quantitative analysis is desired. Since ions are produced with a considerable range of energies, a double focusing instrument is essential.

Figure 33 shows the layout of the Applied Research Laboratories' instrument. A duoplasmatron ion source is used to produce an intense ion beam which is mass resolved before striking the surface of the specimen under investigation. Secondary ions from the target surface are mass analysed in a double focusing mass spectrometer. A scintillation detector enables ion counting methods to be used in the detection system.

It is possible to run the ion source to produce negative ions. This has two advantages. It enables more intense beams to be obtained from electronegative gases such as the halogens. Secondly, it permits analysis of insulators since, with positive ions, insulators charge up and deflect the positive ion beam. With insulators, because the secondary electron coefficient is greater than unity, the surface starts to go positive. This positive potential prevents the escape of the less energetic secondary electrons and a stable situation is quickly achieved.

A very interesting property of the overall instrument is the ability to switch between atomic and molecular spectra by the use of an energy window. It is found that the energy distribution of atomic ions has a much greater high

17 Reed

energy component than is the case for molecular ions. Figure 34 shows a molecular and atomic spectrum obtained for an Al, Mg alloy. This energy window is obtained by varying the secondary acceleration voltage and keeping the electrostatic analyser voltage fixed.

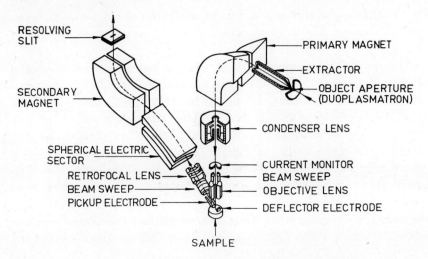

RESOLVING SLIT

PRIMARY MAGNET

EXTRACTOR

OBJECT APERTURE (DUOPLASMATRON)

SECONDARY MAGNET

CONDENSER LENS

SPHERICAL ELECTRIC SECTOR

CURRENT MONITOR

RETROFOCAL LENS — BEAM SWEEP

BEAM SWEEP — OBJECTIVE LENS

PICKUP ELECTRODE — DEFLECTOR ELECTRODE

SAMPLE

Layout of the A.R.L. sputtering microprobe analyser

The atomic spectrum is generally used for analysis purposes, but even the molecular one can help here where overlap occurs, as for example Si^{2+} with N^+.

As an indication of sensitivity the ARL workers claim ion yields in excess of 10^5 counts/sec on the ^{24}Mg isotope in a sample of Al containing 0.03% of Mg.

While not yet achieving the sensitivity of the spark source instrument, the sputter ion type instrument holds considerable promise for the future.

d   Deuterium analysis

Heavy water has an important place in nuclear reactor physics since deuterium is a very efficient moderator of neutrons and has a very much lower neutron capture cross section than hydrogen. In the UK Nuclear Power Programme, one reactor—the Steam Generating Heavy Water Reactor, SGHWR—uses heavy water as a moderator and light water to produce steam. As part of an assessment of the feasibility of heavy water production to sup-

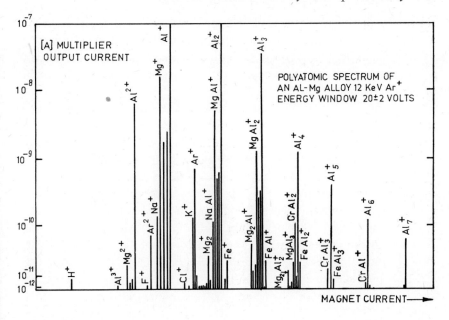

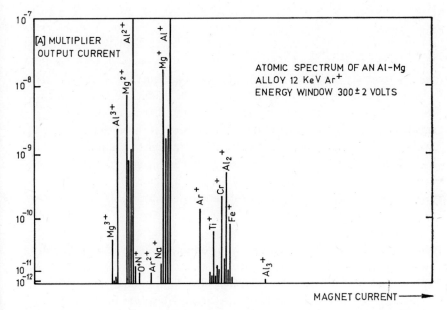

Figure 34   The atomic and molecular secondary ion spectrum obtained from an
Al–Mg alloy with a microprobe analyser

port an installation programme of such reactors the analysis of the deuterium content of feedstocks at potential sites for production plant is necessary. This is important as the capital cost of a production plant, which is about £50,000 per ton of annual output, is approximately inversely proportional to the deuterium content of the feed.

Measurements of very low deuterium concentration in hydrogen have been carried out for a number of years by both single and double collection instruments. A considerable amount of work, particularly on the latter type of instrument, has been carried out by Nief and Botter[66] who have achieved an absolute precision of about 0.1 ppm. This is a field where absolute standards are needed to calibrate the instruments for bias effects arising from probably a wide variety of reasons. These standards exist at the National Bureau of Standards and Vienna. The deuterium content of standards such as SMOW (Standard Mean Ocean Water) is about 158 ppm and for SNOW (literally snow) is about 100 ppm.

A fundamental difficulty exists in low deuterium analysis because of the formation of a molecular ion H_3^+ arising from ion molecule reactions in the ion source. This ion contributes to the HD^+ (mass 3) position where natural level deuterium is measured and its intensity varies as p^2 where p is the hydrogen pressure in the ion source. The contribution may be about 10% at normal source pressures. This difficulty is normally circumvented by plotting mass 3/mass 2 against mass 2 and extrapolating to zero pressure. It is assumed that the H_2^+ intensity at mass 2 is proportional to pressure.

Recently at Aldermaston we have solved the problem in a different way, using the MSG mass spectrometer described earlier in this paper. Here a small water sample is injected into a hot uranium furnace and converted to hydrogen. The hydrogen is leaked into the ultra high vacuum mass spectrometer at a level one hundred times smaller than is normally used. This results in a contribution from H_3^+ that is reduced by a factor 10^4, i.e. to negligible proportions. Measurements are then made by counting HD^+ ions at a rate of about 10^4 ions/sec for fixed times and measuring H_2^+ beam intensities precisely on a plate collector using a digital voltmeter.

This method relies for its success on two factors: firstly, that the scintillation detector counting is nearly absolute, (i.e. that every ion that enters the detector is recorded) and, secondly, that ultra high vacuum conditions exist in the ion source so that there is little or no contribution to the H_2^+ and HD^+ signal from residual gases.

This method of deuterium analysis is still being investigated, but already measurements to better than 1% can be achieved quickly and easily on a routine basis and higher precision should be obtainable.

e Materials problems in controlled thermonuclear reactors

Over the past decade a considerable scientific effort has been devoted to the problem of containing a hot plasma for a sufficiently long time to extract useful power from it. Associated problems arise such as that of producing a suitable vacuum wall material for the reactor. Some of the properties a good wall should have are:

1 a high probability of burying ions that strike it so that they do not return to the plasma as slow particles and cool it by, for example, charge transfer.

2 A low sputtering coefficient to prevent the hydrogen plasma being contaminated by high Z wall atoms that would quickly cool it by Bremsstrahlung radiation losses.

3 Good vacuum properties such as low outgassing rates, etc.

4 Low neutron absorption.

Groups in America, Russia and England are examining this type of problem. At Aldermaston aspects (1) and (2) are being investigated and Figure 35 shows the apparatus being used. An isotope separator produces an intense beam of ions which enters an ultra high vacuum chamber where the particular target under examination is mounted. Various methods are available to monitor what happens when the beam interacts with the surface. One experiment by Freeman *et al.*[67] describes in detail a method to discover the best surface to bury efficiently very large fluxes of fast hydrogen ions. In this work a small residual gas analyser mass spectrometer was used to sample the gases in the target chamber and any change in the trapping coefficient η (where η is the fraction of the bombarding beam which is trapped in the target) would show up as a change in the background hydrogen spectrum. The type of result obtained is shown in Figure 36, which shows the variation of trapping coefficient with target temperature.

Here it can be seen that for titanium, doses of 10^{19} ions/cm^2 can be buried over a wide temperature range with a trapping coefficient of approx. 0.97. The fact that the coefficient is low at low temperatures can be explained by assuming that little or no diffusion of hydrogen into the bulk metal occurs, and the top atomic layers quickly saturate with hydrogen and collisions re-

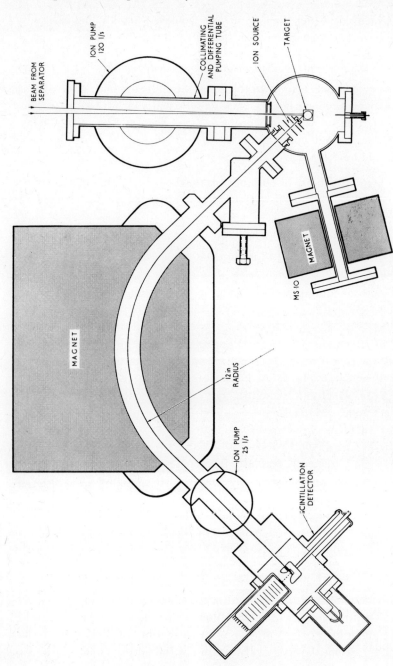

Figure 35 Apparatus for the study of energetic ion—target interactions

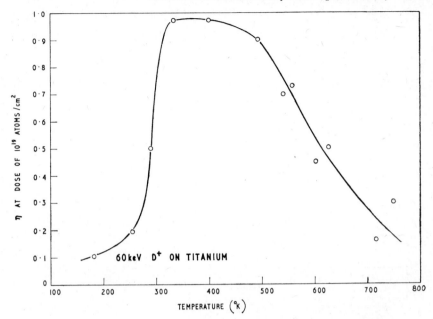

Figure 36 Variation of trapping coefficient with temperature for 60 keV D^+ ions (at a dose of 10^{19} ions/cm^2)

lease it. At high temperatures hydrogen atoms have enough energy to jump the potential barrier at the surface and escape, but at intermediate energies the hydrogen just diffuses into the bulk material. Thus from the ion burial point of view titanium is a very suitable wall material for thermonuclear reactions. Similar results with the reactive metals Nb, Ti, Zr and Er have been found by McCracken *et al.*[68] using the same technique. They showed that the temperature at which the trapping efficiency decreased was linearly related to the heat of formation of the hydride.

Another experiment, again using the same type of apparatus, has been devised to measure ion ranges in solids. Numerous measurements have been performed by Davies *et al.*[69] and his group at Chalk River, using radioactive ions that are implanted in solids. The surface is stripped off electrolytically in layers, and the range found by counting techniques. However, in the method described by Freeman and Latimer[70] energetic deuterium ions are injected into solids and then the surface is sputtered off whith a beam of heavy ions. As the deuterium atoms are released they are measured using the residual gas analyser mass spectrometer. Figure 37 shows the results of a

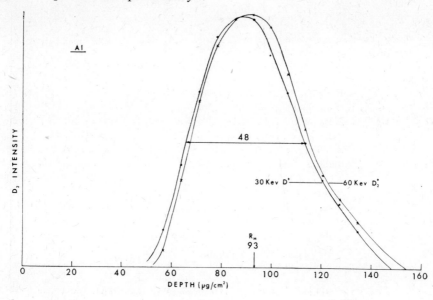

Figure 37   The depth distribution of 30 keV D$^+$ and 60 keV D$_2^+$ ions in aluminium

range measurement of 30 keV D$^+$ and 60 keV D$_2^+$ ions in aluminium. A well-defined range can be measured very quickly by this method.

A 12 inch radius mass spectrometer is also shown in Figure 32 and its ion source "views" the area where the isotope separator beam strikes the target. This is required to examine ions and neutrals leaving the target. The ion source has an electron beam that can ionize neutral particles. This spectrometer has been used to examine the energy distribution of hydrogen ions back scattered from the target surface[71]. It is presently being used to measure the sputtering coefficient for niobium ions and hydrogen ions bombarding a niobium target. Niobium is a material with properties that make it potentially a useful wall material for thermonuclear reactors.

References

1. Thomson, J.J., *Phil. Mag.*, (6), **24**, 209, 668 (1912).
2. Aston, F.W., *Phil. Mag.*, **38**, 709 (1919).
3. Oliphant, M.L., Shire, E.S., and Crowther, B.M., *Proc. Roy. Soc.*, **A 146**, 922 (1934).
4. Stephens, W.E., *Phys. Rev.*, **45**, 513 (1934).
5. Inghram, M.G. and Hayden, R.J., *Nuclear Science Series Report No. 14*, National Academy of Sciences, Washington National Research Council, 1954.

6. Wilson, H.W., Munro, R., Hardy, R.W.D., and Daly, N.R., *Nucl. Instr. and Meth.*, **13**, 269 (1961).

7. Daly, N.R., *Rev. Sci. Instr.*, **31**, 264, 720 (1960).

8. White, F.A., Rourke, F.M., and Sheffield, J.G., *Knolls Atomic Power Lab. Report*, **KAPL-1843**, 1958.

9. Mattauch, J. and Herzog, R., *Z. Phys.*, **89**, 786 (1934).

10. Reynolds, J.H., *Rev. Sci. Instr.*, **27**, 928 (1956).

11. Freeman, N.J., Daly, N.R., and Powell, R.E., *Rev. Sci. Instr.*, **38**, 945 (1967).

12. Nier, A.O., Ney, E.P., and Inghram, M.G., *Rev. Sci. Instr.*, **18**, 294 (1947).

13. Sidenius, G. and Skilbreid, O., *Electromagnetic Separation of Radioactive Isotopes* (edited by Higatsberger, M.J. and Viehböck, F.P.), p. 234, Springer-Verlag, Vienna, 1961.

14. Thode, H.G. and Graham, R.L., *Can. J. Research*, **A25**, 1 (1947).

15. Macnamara, J., Collins, C.B., and Thode, H.G., *Phys. Rev.*, **78**, 129 (1950).

16. Clarke, W.B. and Thode, H.G., *Can. J. Phys.*, **42**, 213 (1964).

17. Dietz, L.A., Pachucki, C.F., and Lord, G.A., *Anal. Chem.*, **35**, 797 (1963).

18. McNair, A. and Wilson, H.W., *Phil. Mag.*, **6**, 563 (1961).

19. Schaeffer, O.A. and Zähringer, J., *Phys. Rev.*, **113**, 674 (1958).

20. Goebel, A. and Zähringer, J., *Z. Naturforsch.*, **16a**, 1042 (1961).

21. Voshage, H. and Hintenberger, H., *Z. Naturforsch.*, **16a**, 1042 (1961).

22. Gradsztajn, E., Ephene, M., and Bernas, R., *Phys. Letters*, **4**, 257 (1963).

23. Forsling, W., Herrlander, C.J., and Ryde, H. (Editors), Proceedings of International Symposium, Lysekil, Sweden 1966. *Arkiv for Fysik*, **36**, (1967).

24. Klapisch, R. and Bernas, R., *Nucl. Instrum. Meth.*, **38**, 291 (1966).

25. Amarel, I., Bernas, R., Chaumont, J., Foucher, R., Gastrzebski, J., Johnson, A., Klapisch, R., and Teillac, J., *Ark. Fys.*, **36**, 77 (1967).

26. Friedlander, G., Friedman, L., Gorden, B.M., and Yaffe, L., *Phys. Rev.*, **129**, 1809 (1963).

27. Amarel, I., Bernas, R., Foucher, R., Gastrzebski, J., Johnson, A., Teillac, J., and Gauvin, H., *Phys. Lett.*, **24B**, 402 (1967).

28. Chaumont, J., Roeckl, E., Nir-el, Y., Thibault-Philippe, C., Klapisch, R., and Bernas, R., *Phys. Letters*, **29B**, 652 (1969).

29. Klapisch, R., Thibault-Philippe, C., Detraz, C., Chaumont, J., and Bernas, R., *Phys. Rev. Letters*, **23**, 652 (1969).

30. Nier, A.O., Booth, E.T., Dunning, J.R., and Grosse, A.V., *Phys. Rev.*, **57**, 546 (1940).

31. Day, G.M., Tucker, A.B., and Talbert, W.L., *Ames Lab. Res. and Dev. Rep.*, **No. IS-1567**, 1967.

32. Alvager, T., Naumann, R., Petry, R.F., Sidenius, G., and Darrah-Thomas, T., *Phys. Rev.*, **167**, 1105 (1968).

33. Quisenberry, K.S., Scolman, T.T., Nier, A.O., *Phys. Rev.*, **102**, 1071 (1956).

34. Inghram, M.G., Brown, H., Patterson, C., and Hess, D.C., *Phys. Rev.*, **80**, 916 (1950).

35. Katcoff, S., *Nucleonics*, **16**, 78 (1958).

36. Thode, H.G. and Graham, R.L., *Can. J. Research*, **A25**, 1 (1947).

37. Wanless, R.K. and Thode, H.G., *Can. J. Phys.*, **33**, 541 (1955).

38. de Laeter, J.R. and Thode, H.G., *Can. J. Phys.*, **47**, 1409 (1969).

39. Tyrrell, A.C., Roberts, J.W., and Ridley, R.G., *J. Sci. Instr.*, **42**, 806 (1965).

40. Third, K.S. and Tomlinson, R.H., *Can. J. Phys.*, **47**, 275 (1969).
41. Meek, M.E. and Rider, B.F., *Report No. APED-5398-A*, Vallecitos, 1968.
42. Srinivasan, B., Alexander, E.C., and Manuel, O.K., *Phys. Rev.*, **179**, 1166 (1968).
43. Schmitt, H.W., Kiker, W.E., and Williams, C.W., *Phys. Rev.*, **137**, 8837 (1965).
44. Nervik, W.E., *Phys. Rev.*, **119**, 1685 (1960).
45. Glendinin, L.E. and Steinberg, E.P., *J. Inorg. and Nucl. Chem.*, **1**, 45 (1957).
46. Eccleshall, D.E., Yates, M.J.L., Cookson, J.A., and Simpson, J.J., *Congrès International de Physique Nucléaire*, Paris, July 2–8, 1964. Simpson, J.J., Eccleshall, D.E. Yates, M.J.L., and Freeman, N.J., *Nuc. Phys.*, **94**, 177 (1967).
47. Brostrøm, K.J., Huus, T., and Koch, J., *Nature*, **160**, 498 (1947).
48. Mileikowsky, C. and Wahling, W., *Phys. Rev.*, **88**, 1254 (1952).
49. Bergström, J., Brown, F., Davies, J.A., Geiger, J.S., Graham, R.L., and Kelly, R., *Nucl. Instr. and Meth.*, **21**, 249 (1963).
50. Smythe, W.R. and Hemmendinger, A., *Phys. Rev.*, **51**, 178, 1052 (1937).
51. Nier, A.O., *Rev. Sci. Instr.*, **11**, 212 (1940); **18**, 398 (1947).
52. Anderson, G., *Phil. Mag.*, **45**, 621 (1954).
53. Uhler, J., Forsling, W., and Astrom, B., *Arkiv. Fysik*, **24**, 421 (1963).
54. Rudstam, G., *Fourth Scandinavian Isotope Separator Symposium, 1963*, Paper 18.
55. Dempster, A.J., *Phys. Rev.*, **71**, 829 (1947).
56. Walker, W.H., *Ph.D. Thesis*, McMaster University, Hamilton, Canada, 1954.
57. White, F.A., Rourke, F.M., Sheffield, J.C., and Dietz, L.A., *I.R.E. Trans. Nucl. Sci.*, **8**, 18 (1961).
58. Forman, L. and White, F.A., *Nucl. Sci. and Engin.*, **28**, 139 (1967).
59. Fenner, N.C. and Large, R.S., *J. Inorg. Chem.*, **29**, 2147 (1967).
60. Farrar, H., Fifteenth Annual Conference on Mass Spectrometry and Allied Topics, Denver, Colorado, page 221, 1967.
61. Freeman, N.J., Barry, J.F., and Campbell, N.L., *J. Nucl. Energy*, in press.
62. Honig, R.E., *J. Appl. Phys.*, **29**, 549 (1958).
63. McHugh, J.A. and Sheffield, J.C., *Appl. Phys.*, **35**, 512 (1964).
64. Herzog, R.F.K., Poschenrieder, W.P., Ruedenauer, F.G., and Satkiewicz, F.G., *Fifteenth Annual Conference on Mass Spectrometry and Allied Topics*, Denver, Colorado, p. 301, 1967.
65. Andersen, C.A., *Int. J. Mass Spect. Ion Phys.*, **2**, 61 (1969).
66. Nief, G. and Botter, R., *Adv. in Mass Spect.*, **1**, 515 (1959).
67. Freeman, N.J., Latimer, I.D., and Daly, N.R., *Nature*, **212**, 1346 (1966).
68. McCracken, G.M., Jefferies, D.K., and Goldsmith, P., *Proc. of 4th Int. Vac. Congress* (Manchester), 149 (1968).
69. Davies, J.A., Friesen, J., and McIntyre, J.D., *Can. J. Chem.*, **38**, 1526 (1960).
70. Freeman, N.J. and Latimer, I.D., *Can. J. Phys.*, **46**, 467 (1968).
71. McCracken, G.M. and Freeman, N.J., *J. Phys. B.*, **2**, 661 (1969).

Industrial applications of mass spectrometry

A. QUAYLE

Shell Research Ltd., Thornton Research Centre,
P.O. Box 1, Chester CH1 3SH, England

INTRODUCTION

Nowadays mass spectrometry finds very wide application in industry and I shall try to illustrate not only the various fields where the technique is used, but also the different kinds of instrument found useful in the particular circumstances surrounding each kind of application. A mass spectrometer is essentially an analytical device, telling us both how much and what kind of material is producing ions in its source. Hence, the emphasis in this paper will be heavily on analysis rather than on the investigation of ionization and dissociation processes. This does not mean that industry does not do mass spectrometric research. Far from it, but the investigations are usually directed towards a sufficient understanding of a phenomenon which is impeding the desired analytical solution.

There are few recorded cases of mass spectrometers being located in actual production plants, analysing the process streams directly, although I will return to this application later. Instead, samples tend to be brought to a mass spectrometer in an analytical laboratory. However, it is likely that research or development applications exceed strictly process analyses in the industrial applications of mass spectrometry; I hope to give due weight to both these fields of application. My approach, then, will be to consider several industries in turn, covering analytical and research applications within each.

MASS SPECTROMETRY
IN THE PETROLEUM INDUSTRY

I hope you will accept my starting with my own industry. Not only am I most familiar with this field, but the petroleum industry can fairly claim to have been the first major user of mass spectrometers outside the nuclear energy world and to have contributed significantly to their development, not only directly, but also by the incentive they have given to instrument manufacturers to extend the performance of their machines. In fact, I have heard the petroleum industry accused, in years gone by, of making such a good job of the mass spectrometric analysis of hydrocarbons as to have delayed the wider application of mass spectrometers to chemistry in general, by creating the impression that they were only useful for hydrocarbon analysis.

First, a digression into the composition of petroleum, which consists of a complex mixture of hydrocarbons and, to a lesser extent, non-hydrocarbons. The boiling range of a typical de-asphaltened crude (Wasson, Texas; see Coleman[1]) extends up to 650°C (corrected to atmospheric pressure), leaving a 10%w residue. The molecular weights of the distillate portions extend up to about 800 and this particular residue is shown, by gel permeation chromatography, to contain materials of molecular weights up to 5300. The asphaltenes already removed would, of course, be even heavier; they comprise about 0.5 to 1 % of a full-range crude, or 3 to 5 % of a "short residue" from a bituminous feedstock. The hydrocarbons within these molecular weight ranges will consist of paraffins (normal and branched), cycloparaffins (single and multi-ringed) and aromatics (single and multi-ringed—condensed and non-condensed, alkyl and cyclo-alkyl substituted or condensed). Unsaturated (olefinic) compounds are not very prevalent in crude oils, but may exist in many forms in refinery streams, especially those which have undergone thermal or catalytic reforming. Among the non-hydrocarbons the sulphur compounds may be considered as derived from the hydrocarbons by substitution of a sulphur atom for a methylene group (CH_2), producing mercaptans, sulphides, disulphides or thiacyclanes, or by substitution of a sulphur atom for a CH group in an aromatic system, producing thiophenes, benzothiophenes etc. Nitrogen compounds can be basic (amines or pyridines) or non-basic (indoles or carbozoles); oxygen compounds are most likely to be furans and benzofurans. These are the main compound types encountered, although many crudes contain considerable amounts of vanadium as porphyrins. When the number of possible isomers of any particular compound is considered, the problems facing the mass spectroscopist in analysing this brew can be appreciated.

Gas analyses

In refining operations, the problem is usually that of analysing particular streams. Some companies have diverted this problem from mass spectrometry to gas chromatography, but others have put effort into increasing the efficiency of the operation by applying computers to mass spectrometry. Simple gas analysis depends on a knowledge of the components in the gas mixture, enabling the mass spectrometer to be calibrated with reference compounds, to determine their cracking patterns and relative sensitivities. Skilful operation can keep the instruments in a constant operating condition, enabling these calibrations to be used for several months. The intensity of a peak h_i in the mass spectrum of the gas mixture at m/e "i" is given by:

$$h_i = s_{i1}p_1 + s_{i2}p_2 + \cdots = \Sigma \, s_{ij}p_j$$

where s_{ij} is the sensitivity coefficient of component j at m/e i, and p_j is the partial pressure of component j in the mixture. The observed mass spectrum may be written as the column vector H, equal to the product of the column vector of partial pressure P multiplied by the matrix of the sensitivity coefficients S. The desired partial pressures are obtained by multiplying the mass spectrum (or sufficient peaks selected therefrom, equal to the number of components) by the inverse matrix of the sensitivity coefficients of the component gases, obtained during calibration:

$$\mathbf{P} = S^{-1}\mathbf{H}$$

The composition of the gas in mole per cent (equal to volume per cent and which the computer can readily convert into weight per cent) is given by:

$$c_j = p_j/\Sigma p_j$$

where c_j is the concentration (in mole per cent) of the jth component. Again, the art of mass spectrometry lies in selecting peaks to represent the various components so that the sensitivity coefficient matrix is well-conditioned and inverts readily. An example of a typical gas analysis is given in Figure 1. The peaks used for this analysis occurred at the mass numbers listed below.

Peaks used for cracked gas analysis (major diagonal of calculation matrix):

	m/e		m/e		m/e
hydrogen	2	carbon dioxide	44	i-pentane	57
methane	16	propane	29	total pentenes	70
nitrogen	28	total butenes	56	cyclo-pentene ⎫	
ethylene	26	i-butane	43	cyclo-hexene ⎬	67
ethane	30	n-butane	58	total C_6	84
propylene	42	n-pentane	72	total C_7	98

Three stages of development have occurred, principally in America, in this form of analysis. First, peaks were measured by hand; then they were measured by a relatively slow digitizing device (the "Spectro-Sadic"[2]) and later by fast digitizing equipment (the "Mascot"[2] and more recent and efficient developments of this kind[3-8]). The most recent forms of computer

	% molar		% molar
Hydrogen	13.4	Total butenes	6.4
Methane	4.8	Isobutane	6.7
Water	a	n-Butane	1.4
Nitrogen	45.8	Isopentane	3.2
Ethylene	1.8	n-Pentane	0.1
Ethane	2.1	Total pentenes	2.2
Oxygen	a	Cyclopentene ⎫	
Propene	7.2	Cyclohexene ⎬	0.2
Carbon dioxide	a	Total C_6	0.5
Propane	3.6	Total C_7	0.1

a Reported on a water-, air- and carbon dioxide-free basis

Figure 1   Cracked gas analysis

control can interact with the mass spectrometer. The magnetic field is held extremely constant by, say, a Hall-effect probe and control system. The computer calculates the ion accelerating voltage needed to focus an ion of certain mass and applies this voltage to the source. The applied voltage is sampled, compared with the required value and corrected as necessary. When the computer senses that electric and magnetic fields are correct, the focused ion beam is sampled, digitized and stored in the memory. The computer repeats the operation for the next mass number desired, re-adjusts the ion accelerating voltage, stores the ion beam intensity and steps its way through the selected peaks in the spectrum. Having obtained all the necessary peaks, it quickly calculates the composition from the calibrations stored in its memory, prints out "*n*" copies of the analysis and instructs the operator to change the sample. Over 60 samples per eight-hour shift can be analysed in this way.

Fraction boiling range component	B.P. °F	100–140	140–165	165–187	187–205	205–215	215–225	°F
n-Pentane	96.9	+						
Cyclopentane	120.7	+						
2,2-Dimethylbutane	121.5	+						
2,3-Dimethylbutane	136.4	+	+					
2-Methylpentane	140.5	+	+					
3-Methylpentane	145.9	+	+					
n-Hexane	155.7	+	+	+				
Methylcyclopentane	161.3	+	+	+				
2,2-Dimethylpentane	174.5	+	+	+				
2,4-Dimethylpentane	176.9		+	+				
Cyclohexane	177.3		+	+	+			
3,3-Dimethylpentane	186.9			+	+			
1,1-Dimethylcyclopentane	190.1			+	+			
2,3-Dimethylpentane	193.6			+	+			
2-Methylhexane	194.1			+	+			
Cis and trans-1,3-di MecyC$_5$	195–197			+	+			
3-Methylhexane	197.3				+			
Trans-1,2-di MecyC$_5$	197.4			+	{+			
Cis-1,2-di MecyC$_5$	211.2					+	+	
3-Ethylpentane	200.2				+	+		
n-Heptane	209.2				+	+	+	
2,2,4-Trimethylpentane	210.6				+			
Methylcyclohexane	213.7				+	+	+	
Ethylcyclopentane	218.2					+	+	
1,1,3-tri MecyC$_5$	220.8					+	+	
2,2-Dimethylhexane	224.3						+	
2,5-Dimethylhexane	228.4						+	
1,Trans-2, cis-4-tri MecyC$_5$	228.7						+	
2,4-Dimethylhexane	229.0						+	
1, Trans-2, cis-3-tri MecyC$_5$	230.4						+	
3,3-Dimethylhexane	233.5						+	
Benzene	176.2	+	+	+	+	+	+	
Toluene	231.1	+	+	+	+	+	+	

Figure 2 Gasoline component analyses

Hydrocarbon type analyses

Because the number of possible compounds increases rapidly with molecular weight, we soon reach the stage where it is impossible to analyse for single compounds. In the gasoline range, we require narrow distillation fractions to carry out single-component analyses successfully (Figure 2). For wider range gasolines and heavier materials, hydrocarbon-type analyses are applied. The basis of this approach is that, whilst individual hydrocarbons differ in detail, similar compound types have similar mass spectra. For example, Figure 3 shows the similarity of some paraffin spectra and, based on this sort of knowledge, certain peaks are used to estimate a class of hydrocarbons. Thus, the sum of the intensities of the peaks at m/e 43, 57, 71 and 85 may be taken as representative of the paraffins; the sum of 41, 55, 69 and 83 of cycloparaffins (mono-naphthenes) plus mono-olefins (if present); Σ 53, 67, 81 may represent dinaphthenes plus dienes and acetylenes; Σ 77, 78, 91, 92, 105 and 106 may be used for alkylbenzenes, and so on. Different experimenters will differ as to the particular groupings they use (where to terminate the series or whether to include, say, m/e 78 in the alkylbenzenes) and as to the relative weight they give to each series, but the approaches are essentially the same. Figure 4 shows the results of an analysis carried out this way.

I do not wish to bore you with excessive details of hydrocarbon analyses, but there are four further topics I want to touch upon before leaving the subject. The first concerns a compound-classification system, called the "z-number grid", or, sometimes, rectilinear array.

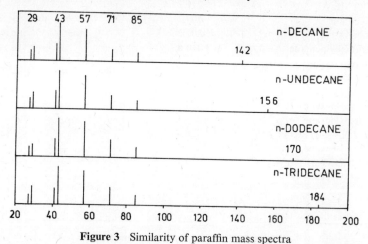

Figure 3   Similarity of paraffin mass spectra

Hydrocarbon type		Mass spectrometry	Chromato- graphy	Difference
Paraffins	C_nH_{2n+2}	10.7	13.7	-3.0
Mono-olefins + mono-naphthenes	C_nH_{2n}	39.6	41.9	-2.3
Dienes, di-naphthenes cyclenes + acetylenes	C_nH_{2n-2}	15.9	17.6	-1.7
Further unsaturateds (e.g. cyclic dienes)	C_nH_{2n-4}	2.6	–	$+2.6$
Alkyl benzenes	C_nH_{2n-6}	25.7		
Styrenes + tetralins	C_nH_{2n-8}	4.9	26.9	$+4.3$
Naphthalenes	C_nH_{2n-12}	0.6		

Figure 4   Comparison of gasoline analyses (in volume %)

z-number grid

If we express a hydrocarbon by the general formula C_nH_{2n+z}, the value of "z" can define the hydrocarbon type, e.g. if z is $+2$, the compound is a paraffin; if $z = -6$, an alkylbenzene; if $z = -14$, a diphenyl—and so on. We can construct a table of mass numbers, with 14 columns, each headed by a particular z number, and the molecular weight of a hydrocarbon will fall in a column headed by the z number of its hydrocarbon type (Figure 5). Many hydrocarbons will have significant parent peaks (especially if they are aromatics) and much of the ion intensity is associated with the breaking of alkyl (perhaps substituent) chains to give major fragments in the column containing m/e $M-1$ (M being the molecular weight). Thus, if we tabulate a mass spectrum on a grid of this form, and sum the ion intensity in each

C No.	z No.													
	-11	-10	-9	-8	-7	-6	-5	-4	-3	-2	-1	0	$+1$	$+2$
5	59	60	61	62	63	64	65	66	67	68	69	70	71	72
6	73	74	75	76	77	78	79	80	81	82	83	84	85	86
7	87	88	89	90	91	92	93	94	95	96	97	98	99	100
8	101	102	103	104	105	106	107	108	109	110	111	112	113	114
9	115	116	117	118	119	120	121	122	123	124	125	126	127	128
10	129	130	131	132	133	134	135	136	137	138	139	140	141	142

Figure 5   Part of z-number grid

column, we can often get a very good idea of the nature of an unknown compound from the two columns containing most of the ion intensity, or detect whether we probably have a mixture. This technique is not infallible, but it can often be of great assistance. The concept can be extended to non-hydrocarbons by remembering that we can replace a methylene group (CH_2: 14 mass units) by an oxygen atom (O: 16 units) increasing the z-number by two. Thus, alcohols, formally derived from the alkanes, appear in the $z = +4$ column and phenols in the $z = -4$ column (2 greater than the "related" alkylbenzenes). Similar arguments can be applied to nitrogen compounds (amines turn up as $z = +3$) and sulphur compounds (sulphides arise in the $z = +6$ column). One must always be careful of the overlap of 14 mass units when the degree of unsaturation is large or there are hetero-atoms present. I will clarify this statement by pointing out that nonane (C_9H_{20}) and naphthalene ($C_{10}H_8$) both have molecular weight 128 and will therefore appear in the same z-number column on the grid, although $z = +2$ for nonane and $z = -12$ for naphthalene. Another problem arises with benzothiophene which can be "formally" derived from indene ($z = -10$), replacement of CH_2 by S causing z to increase by 4, giving $z = -6$ for benzothiophene. But the alkylbenzenes fall on the -6 series and difficulties can arise in a single-focusing mass spectrometer in distinguishing between the two compound types. This leads me to my second related topic.

High resolving power mass spectrometry in petroleum analyses

I have already mentioned hydrocarbon-type analyses (for gasolines) where groups of peaks are selected as representative of different hydrocarbon types. To resolve the difficulties that arise when two types have the same nominal mass, high resolving power can be used to detect these different molecules because of the slight mass difference between them. The alkylbenzenes and benzothiophenes of the previous example differ by 90.5 m.m.u. and are separable, at a resolving power of 10,000, up to m/e 905. Gallegos *et al.*[9] have, used this fact to produce a method of analysing for 19 hydrocarbon types 7 saturated, (0–6 rings/molecule), 9 aromatic (1–4 rings) and 3 aromatic sulphur types, for distillate fractions in the 500–950°F (160–510°C) boiling range, i.e. in the gas oil and lubricating oil ranges. They introduced 2 μl oil into the all-glass heated inlet of an AEI MS9, at 300°C. Scanning took 45 minutes, which was later reduced to 5 minutes. As typical of the paraffins, they used m/e's 71, 85, 99 and 113; for the alkylbenzenes, m/e's 91, 105, 119 and 113 and for the benzothiophenes m/e's 147, 161 and 175; but for this last class of compounds the second component of the multiplet was used as the operative peak. Figure 6 illustrates some of the multiplets observed in these oil fractions.

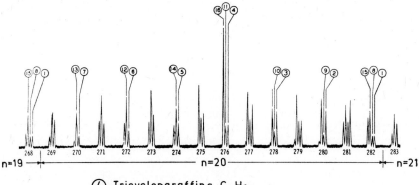

④ Tricycloparaffins C_nH_{2n-4}

⑪ Phenanthrenes C_nH_{2n-18}

⑯ Naphthobenzothiophenes $C_nH_{2n-22}S$

Figure 6 Segment from a high-resolution mass spectrum of the 800–950°F fraction of an Arabian crude (from reference 9)

Low voltage spectra

The third topic I wish to mention concerns the analysis of compounds in the kerosine and light gas oil ranges (C_{12} to C_{20})[10]. One often requires detailed compositional information in this region and the mass spectroscopist brings forward yet another of the modes of operation of his instrument to help solve his problem. You will recall the general shape of the ionization efficiency curve. In the low voltage region, two factors are of importance: (1) the ionization potentials (I.P.'s) of unsaturated compounds are lower than those of saturated ones; the I.P.'s of olefines are lower than those of alkanes and of aromatics lower than of olefines; (2) the appearance potentials of fragment ions are, in general, higher than those of molecular ions: extra energy is required to break the bonds. Advantage is taken of these two factors in simplifying mass spectra. Reducing the voltage below the appearance potentials of saturated hydrocarbons simplifies the spectrum of a complex mixture. The saturates are no longer visible and the overlap noted previously between naphthalene and nonane is no longer observed. Reducing the voltage still further permits us to observe only the parent peaks of the aromatic molecules, without interference from the fragment ions; the spectrum becomes much simpler to interpret. A word of warning is in order here. Do not be disturbed by the apparently ridiculously low voltage which must be used. The indicated electron volts may need to be well below the known ionization potentials of the molecules before fragmentation ceases. This happens for two reasons: (1) the electrons from the heated filament have a fairly wide energy spread and (2) contact potentials in the source can make the effective potentials significantly different from the applied potentials as indicated by the meters.

Figure 7 Composition of thermally stable and unstable jet fuels (extracted from reference 10)

	Stable	Unstable
Aromatic part only:		
Benzenes, %v	5.7	9.0
Indanes, %v	0.5	2.6
Indenes, %v	–	0.3
Naphthalenes, %v	–	3.0
Acenaphthenes, %v	–	0.1

Although this technique is very useful, it is difficult to apply quantitatively. One is operating on a steep part of the ionization efficiency curve and a slight change in working voltage will alter the sensitivity. Since all the compound curves are not strictly parallel, voltage changes will alter not only the absolute sensitivities but also the relative sensitivities, and an analysis based on one set of relative sensitivities can be thrown into error by a voltage change. It is no use working to a constant meter voltage because, as mentioned earlier, this has only a distant relationship to the effective voltage. The best approach is to use a pair of internal standards and adjust the voltage until a predetermined ratio between their intensities is obtained. An example of this type of analysis is given in Figure 7. A more recent example of the application of low voltage, high resolution mass spectrometry concerns the identification of nitrogen and oxygen compound types in petroleum[10a]. Compounds such as

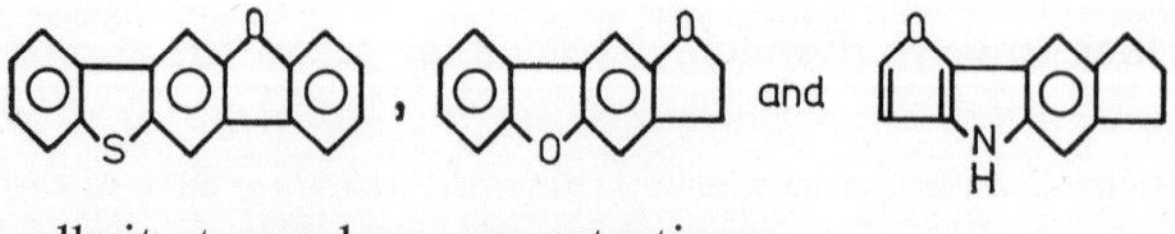

were found, albeit at very low concentrations.

Wax analyses—conventional

Petroleum waxes consist primarily of a mixture of normal paraffins, with lesser amounts of iso-paraffins, mono- and di-cycloparaffins and alkylbenzenes. Figure 8 shows the parent peak region of a typical refined paraffin wax. It is difficult to see the parent peaks, to get a satisfactory molecular weight distribution or to determine the cyclic content. Apart from their generally low intensity the n-paraffins fragment to quite large peaks on the $2n + 1$ series. Thus n-$C_{30}H_{62}$ will give a peak at m/e 351 due to $C_{25}H_{51}^{+}$ which is $1\frac{1}{2}$ times the parent intensity. Now this 351 peak will have associated with it at m/e 352 a heavy-isotope peak due to ^{13}C, which, in a single-focusing instrument, will interfere with the parent peak at m/e 352 of n-$C_{25}H_{52}$, creating the impression that n-$C_{30}H_{62}$ contains 40% n-$C_{25}H_{52}$. Removal of these interferences requires laborious "de-isotoping" calculations, to produce peaks on the $z = +2$ series due to the normal paraffins alone (the iso-paraffins have only small parent peaks and are present in small concentrations, so their contribution to the $z = +2$ series can be neglected). The normal paraffin spectra are quite well known and they can then be calculated out of the spectrum. What is left on the $z = +1$ series can be used to determine the iso-paraffins. Again, after correction, the $z = 0$ series can be used for the mono-cyclanes and the $z = -2$ series for the di-cyclanes (if any). The alkylbenzenes are found from the $z = -6$ series. Care must be taken not to crack the sample by having the inlet system too hot, otherwise the apparent alkylbenzene content will be excessively high. All this involves a lot of arithmetic, although the task can be considerably eased by using modern digital computers. A further cautionary note is in order: be careful about using published spectra, cracking patterns or other coefficients for use in wax analyses. They may not be applicable to your instrument. One can modify cracking patterns by altering the source temperature but one cannot alter the different discrimination characteristics of different designs of mass spectrometer. It has been found impossible, for example, to repeat some critical calibrations, originally obtained on a 180° machine with electric scanning, on a 90° sector instrument with magnetic scanning[11]. The calibrations in question required the m/e 127/226 ratio of n-hexadecane to be 1.45 ± 0.02 and the m/e 127/204 ratio of n-octacosane to be 3.75 ± 0.15.

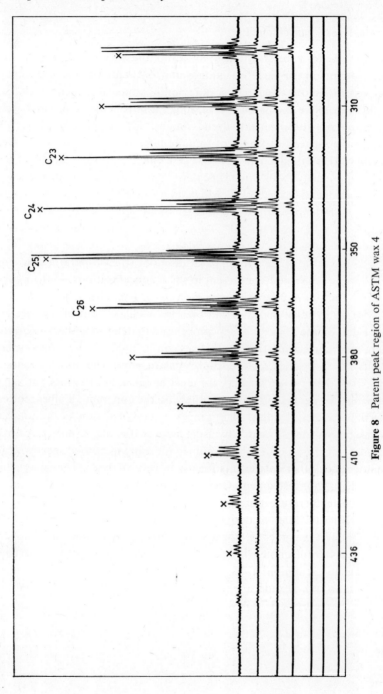

Figure 8 Parent peak region of ASTM wax 4

Wax analyses—field ionization

Field ionization promises to be a technique capable of surmounting many of these problems, as shown by Mead[12]. He obtained good reproducibilty using a razor blade as the field ion source, but far more important was the fact that essentially only parent ions were obtained and the relative sensitivities of the n-paraffin parents were constant over the C_{20} to C_{40} range. Thus much calculation time can be saved, and a lot of useful information can be obtained from a first, visual examination of the spectrum, as shown in Figure 9.

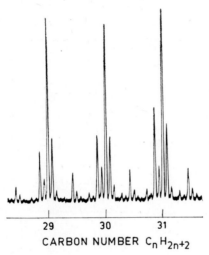

29   30   31

CARBON NUMBER $C_n H_{2n+2}$

Figure 9   Part of field ionization mass spectrum of ASTM wax 1 (from reference 12)

Additives used in petroleum products

Rather than go into details of other lubricating oil hydrocarbon-type and nitrogen-type analyses, I shall mention the use of mass spectrometry in identifying some of the additives used in petroleum products. Figure 10 shows part of the mass spectrum of an aviation turbine oil, the phenyl-naphthylamine antioxidant present being apparent at m/e 219. Aryl phosphates, removed from gasolines by percolation over silica gel, are readily typified by mass spectrometry (Figure 11). Although these spectra can be obtained on a low resolving power instrument, a high resolving power machine, with probe insertion lock, is preferable for identifying organo-metallic additives, as shown in Figure 12.

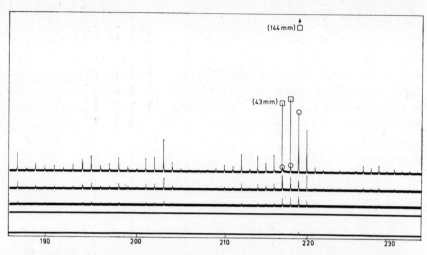

Figure 10 Antioxidant (phenylnaphthylamine, M.Wt. 219) in aviation turbine oil

Phosphate type	Structure of rearranged ion	Mass number
Tri-alkyl	H O $\|$ $HO—P—OH$ $\|$ O H	99
Dialkyl phenyl	H O $\|$ $HO—P—OH$ $\|$ O phenyl	175
Alkyl diphenyl	H O $\|$ phenyl—O—P—O—phenyl $\|$ O H	251

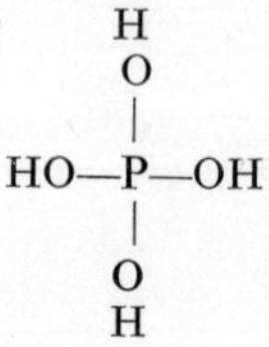

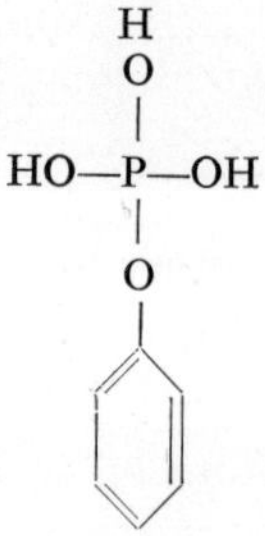

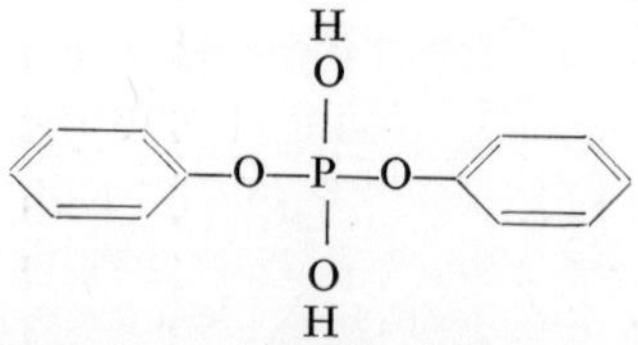

Figure 11 Phosphate types

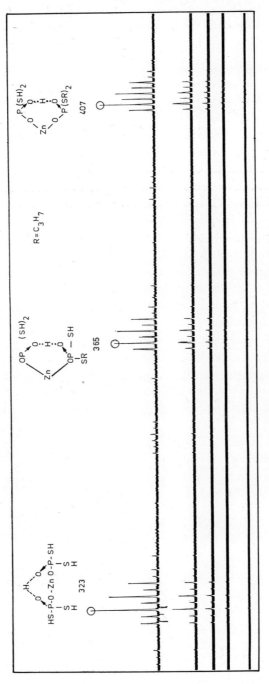

Figure 12   Lubricating oil additive: zinc dialkyldithiophosphate

Research applications in the petroleum industry

So far I have mentioned fairly conventional mass spectrometers, of low and high resolving powers. I will now describe an application involving a combination of gas-liquid chromatography (GLC) and mass spectrometry (MS). As part of a study of the reaction kinetics of a combustion process, it was desired to identify the products of combustion of 2-methylpentane[13,14]. These were trapped in a glass apparatus, and then examined by capillary GLC/MS, leading to the identification of about 65 components. The GLC used helium carrier gas with a pre-column splitter, rejecting about 99% of the charge, and the effluent was split about 9 : 1 between a flame ionization detector and an MS 1D (prototype MS 12) from AEI; no molecular separator was used in the interface. Figure 13 shows the mass spectrum of methyl formate and represents about 1 ng flowing into the mass spectrometer, emerging in a peak 10 seconds wide. Its identity was confirmed by comparison with an authentic calibration sample. The MS 1D's scan rate was 4 to 6 sec per decade, so the range of m/e 15 to 70 shown in Figure 13 took about 3.5 sec. Other compounds found included 2,2,4-trimethyl-oxetane, 2,4-dimethyltetrahydrofuran, 2-methylbutan-3-one and 2-methyl-2,3-epoxybutane.

As part of our research into combustion kinetics, we wish to examine the reactions of compounds on a molecular scale. To this end Dr. J. L. Rosenfeld and his colleagues have built the apparatus shown in Figure 14, to study ion-molecule reactions. An EAI Quad 150 quadrupole mass spectrometer will direct a beam of ions of predetermined low energy across a beam of neutral molecules emanating from the neutral source shown. The product ions will be detected by a second quadrupole mass spectrometer (an EAI Quad 250), which will determine their identity, energy and spatial distribution, since this detecting spectrometer can be moved in three dimensions about the scattering centre.

APPLICATIONS IN THE AGRICULTURAL CHEMICAL INDUSTRY

My colleagues at Woodstock Agricultural Research Centre use an EAI Quad 250 quadrupole mass spectrometer as a specific detector on a GLC in identifying low-level impurities ($\sim 1\%$) in technical pesticide preparations. Figure 15 shows some repeat spectra of the insecticide "Vapona" obtained in this way.

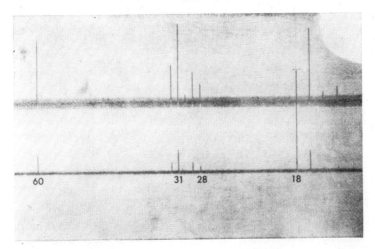

Figure 13 Mass spectrum of methyl formate

Figure 14 Crossed molecular beam apparatus

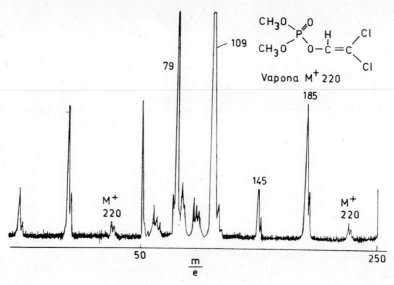

Figure 15   Mass spectrum of "Vapona"

The mass spectrum of the insecticide "Fenitrothion"[15,16] [dimethyl (3-methyl-4-nitrophenyl)-thionophosphate, $(CH_3)_2$ $PS \cdot O \cdot Ph(CH_3)NO_2$], is interesting because it displays a peak at m/e 109 as well as at m/e 125 (Figure 16). What is apparently happening is that rearrangement between the sulphur and oxygen atoms is occurring, thus:

Damico *et al.* have also made use of field ionization mass spectrometry in pesticide analysis[17].

Toxicology is a necessary part of the pesticide business and Tunstall Laboratory of Shell Research Ltd. has collaborated with the Nature Conservancy in analysing the eggs of sea-birds for traces of pesticides. This work has thrown up an interesting and instructive warning. Because electron-capture detectors are very sensitive and fairly specific for halogen com-

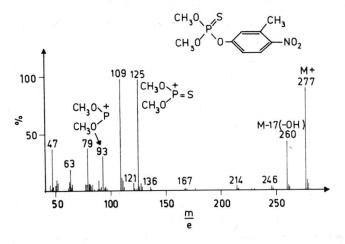

Figure 16   Mass spectrum of "Fenitrothion" (from reference 15)

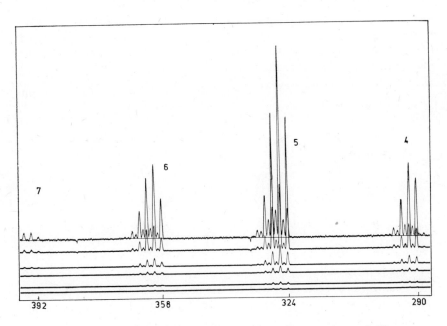

Figure 17   Mixed chlorinated diphenyls (mass spectrum taken at 9 eV)

pounds they are often used as GLC detectors in the analysis of pesticide resi-dues[17a]. The detection of a halogenated compound at the correct retention time for a pesticide (DDT or aldrin) or its metabolites is often taken as evi-dence for the presence of the pesticide, although there is some general awareness of the dangers of misinterpretation. Attachment of a mass spectro-meter (an AEI MS 12) to the GLC showed in these experiments that in some instances the chloro-compounds in the birds eggs were not pesticides but chlorinated aromatic hydrocarbons that are used as plasticizers and in other applications. This illustrates the value of mass spectrometry as an identifying detector in GLC. The mass spectrum of a polychlorinated di-phenyl is seen in Figure 17. Similar observations have been reported by Koeman et al.[17b], who coupled a Varian Aerograph GLC to an Atlas CH 4 mass spectrometer. The chlorinated aromatics they reported included Pheno-chlor DP 6, Arochlor 1260 and Clophen A 60.

Although the work is not strictly related to the agricultural chemical in-dustry, mass spectrometry finds wide use in related fundamental biochemical studies. The method by which enzymes build complex molecules from iso-prenoid units has been studied by examining the farnesol produced from deutero-mevalonic acid, as shown in Figure 18. Elegant proofs of the course of the reactions have been based on whether the enzyme does or does not incorporate deuterium into the resultant farnesol[18].

Scheme

Mevalonic acid: 3,5-dihydroxȳ-3-methylpentanoic acid.
d-labelled acids:

$4S - \left[4D_1 \right]$ — mevalonate　　　　$4R - \left[4D_1 \right]$ — mevalonate

ordinary farnesol　　　　　　　d_3-farnesol

A naturally-occurring plant growth control hormone, abscisic acid, which can be detected in a wide range of leaves, fruit etc., owed its structural

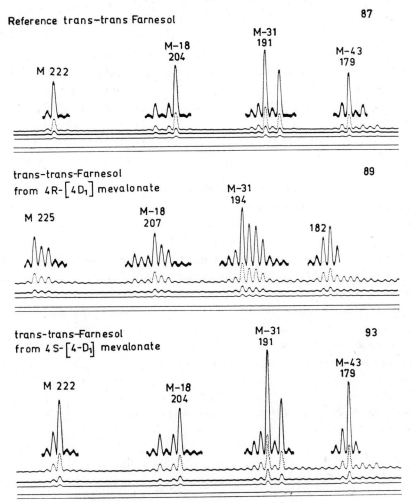

Figure 18 Mass spectrum of Farnesol samples

identification to mass spectrometry, NMR and absorption spectroscopy[19,20]. I am indebted to Mr. V. P. Williams and Dr. G. Ryback of Shell Research Ltd., Milstead Laboratory, for the next figures, illustrating the mass spectrum of abscisic acid (Figure 19) and its modes of fragmentation (Figure 20). These are based on high resolving power mass spectrometry, to identify the elemental compositions of the major peaks, and on the occurrence of metastables, indicating the fragmentation pathways.

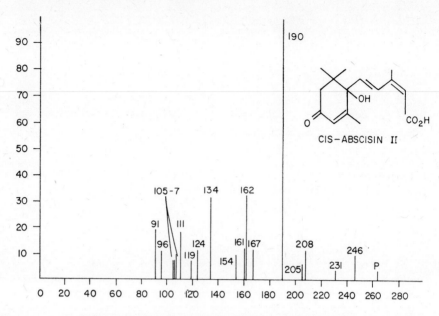

Figure 19   Mass spectrum of abscisin II (abscisic acid)

APPLICATIONS IN THE CHEMICAL INDUSTRY

The workers of Dow Chemicals have made some notable contributions to mass spectrometry following the early lead given by F. W. McLafferty. An interesting application has been the use of mass spectrometry in conjunction with differential thermal analysis[21] (DTA), where the mass spectrometer monitored specific peaks (e.g. H_2O, 18) to identify phase changes detected by the DTA equipment. As an extension to this, Gohlke monitored continuously the mass spectrum of a compound as he heated it in a crucible located in the source region of a Bendix time-of-flight mass spectrometer[22]. The formation of an oxide, with elimination of water, from triphenyltin hydroxide was readily demonstrated ($Ph_3SnOH \rightarrow Ph_3SnOSnPh_3$).

Time-of-flight mass spectrometers have proved especially useful where it has been desired to carry out tests and monitor reactions of this kind. The open geometry of their sources makes them eminently suitable for such work. They have also been used in many laboratories in combination with gas chromatographs, where their fast response made them, at one time, the preferred form of mass spectrometer. Figure 21 shows a very early applica-

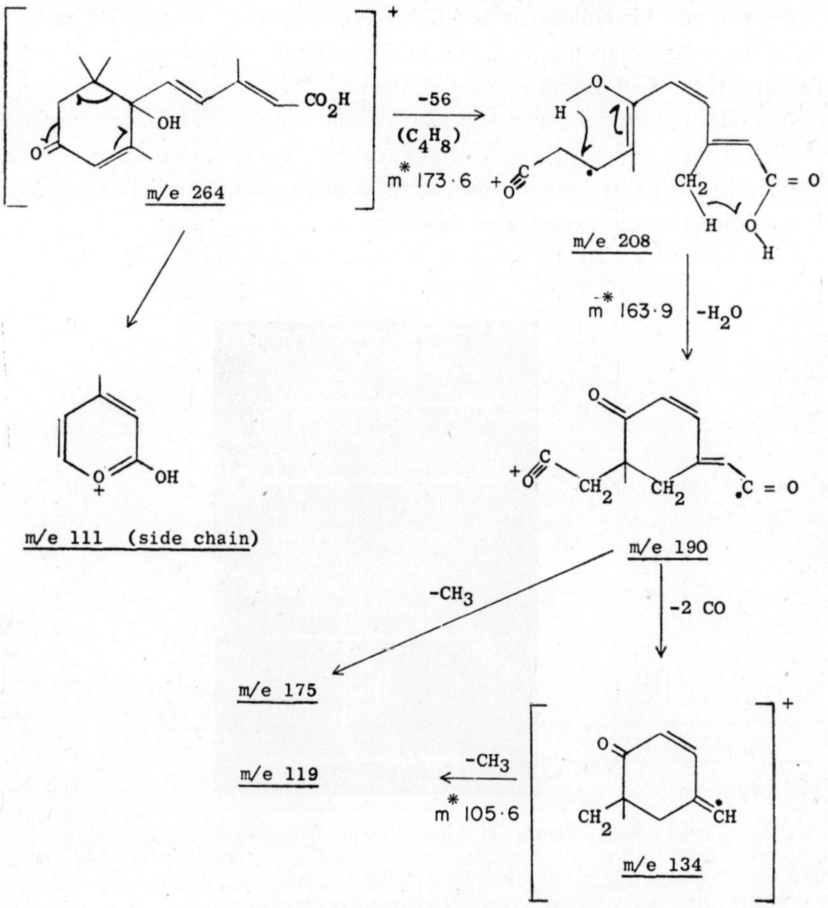

Figure 20 Abscisin fragmentation

tion, from the laboratory of Kodak Ltd., Rochester, New York, in 1960, to the identification of solvents separated by GLC. At that time it was necessary to photograph an oscilloscope screen to obtain the spectrum, but the development of so-called "analogue" detection equipment has since made it possible to record mass spectra in a more conventional, if slower, way.

In similar fashion, an LKB 9000 mass spectrometer/GLC combination has been used in the rubber industry to study the pyrolysis products of isoprene[23].

19 Reed

Beynon and his colleageus at I.C.I. have studied the ionization and dissociation of nitro-aromatics[24] and used the knowledge so gained to derive the structure of an unknown bromonitronaphthalene[25].

A modern chemical process has led to the production of a range of synthetic tertiary mono-carboxylic acids, called "Versatic" acids, which find application in paint formulations, in resin manufacture and elsewhere. Rol has studied the mass spectra of these and other acids[26], and made the surprising discovery that, although m/e 60 in the spectra of primary acids is

Figure 21　Identification of solvents

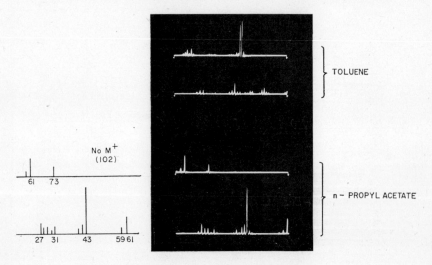

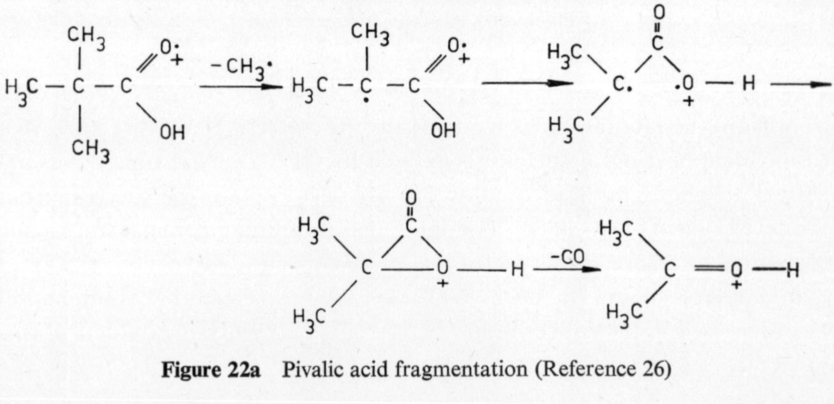

Figure 22a　Pivalic acid fragmentation (Reference 26)

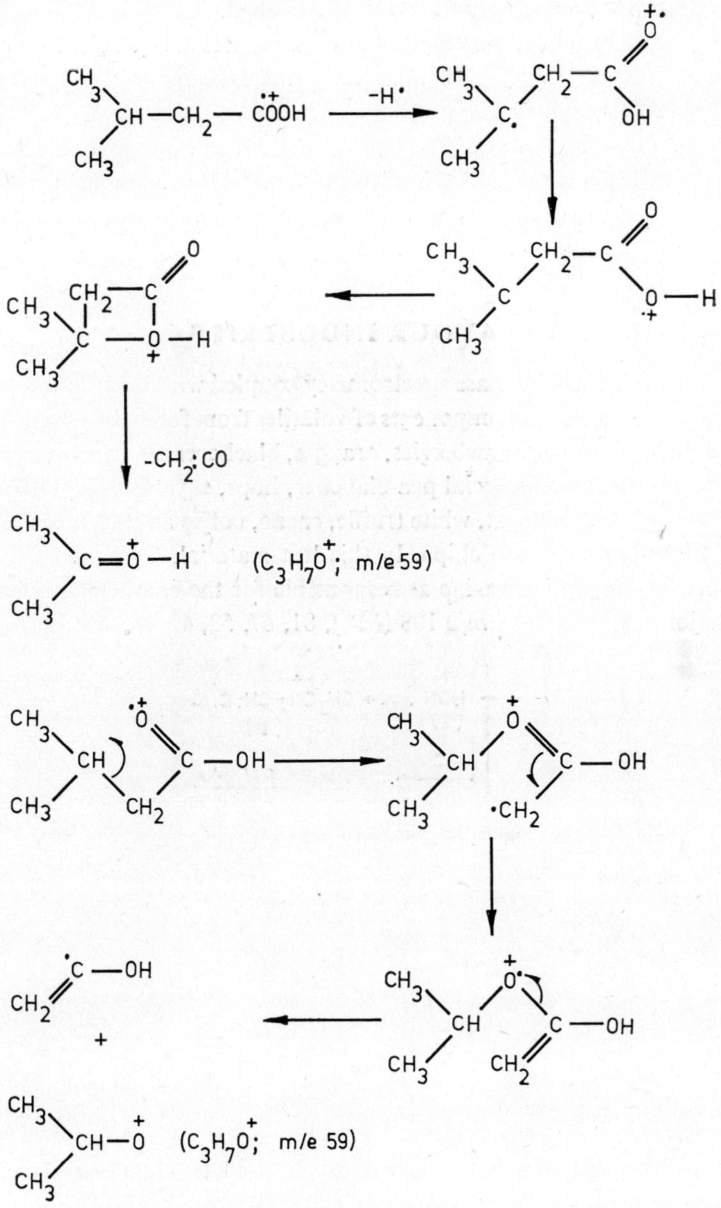

Figure 22b Possible fragmentations of 3-methylbutanoic acid

always $C_2H_4O_2^+$, m/e 59 is not $C_2H_3O_2^+$. Instead, m/e 59 was found to be $C_3H_7O^+$. For example, 3-methylbutanoic acid, $CH_3CH(CH_3)CH_2COOH$, apparently forms $(CH_3)_2COH^+$, instead of the expected CH_2COOH^+. The reaction Rol postulated for pivalic acid (a tertiary acid) is shown in Figure 22(a). Possible routes to $C_3H_7O^+$ (m/e 59) in 3-methylbutanoic acid are given in Figure 22(b). Evidence from metastable transitions should help to resolve these two alternatives. These mass spectrometric investigations were carried out on an AEI MS9.

THE FOOD AND FLAVOUR INDUSTRIES

Extensive use is made of mass spectrometry coupled with GLC in the identification of characteristic components of volatiles from foodstuffs. Substances studied include onions, strawberries, oranges, blackcurrants, apples, cheddar and blue cheese, a commercial pea blancher, hops, soyabean oil, citrus leaf oils, cod fish, chicken meat, white truffle, cacao, coffee and, of special interest to Englishmen, potato chips. In this last material, Deck and Chang[27] identified 2,5-dimethylpyrazine as responsible for the characteristic odour. The major peaks occur at m/e 108 (M^+), 81, 67, 52, 42, 39, **27**:

or

Because of their inherently fast response, time-of-flight mass spectrometers were used almost exclusively in the early applications of GLC/MS. Nowadays, of course, magnetic deflection (single and double-focusing) and quadrupole instruments are used as well. McFadden and Teranishi[28] described some of their original work on fruit volatiles using a Bendix Model 12 T.O.F. mass spectrometer. This had the advantage that they were able to get a form of chromatogram by recording the m/e 15 peak continuously, as received by one of the mass spectrometer "gates". The other "gate" was monitored on an oscilloscope and when an interesting spectrum appeared, it was recorded on a high-speed oscillograph, covering m/e 24 to 600 in 6 seconds.

GAS ANALYSES

In the metallurgical industry

Aspinal[29] has described the use of a small mass spectrometer (AEI MS10) to analyse gases in steels, molybdenum and zirconium by the vacuum fusion technique. He found oxygen determinations at the 116 p.p.m. level reproducible to ± 6 p.p.m. and nitrogen at the 31 p.p.m. level to ± 3 p.p.m. His limits of detection were 0.1 µg for oxygen and nitrogen and 0.01 µg for hydrogen.

In the pure-gas industry

In experiments concerned with the analysis of high-purity helium and hydrogen, Parkinson[30] found it possible to operate an MS2 (G) mass spectrometer satisfactorily at a reservoir pressure of 70 mm, this pressure being used to accentuate the minor components. To reduce background effects, he decided to use a dynamic sampling system, with his gas flowing continually past (and through) the mass spectrometer. By reducing the leak rate by a factor of 10, he could operate at an upstream pressure in the pipelines of 800 mm, which gave adequate sensitivity at the part-per-million level. Naturally, his major component peaks were off-scale, and, although he was able to use m/e 1 for hydrogen, he had to use the ion gauge indicated pressure, after calibration, for helium.

Roboz[31] modified the inlet system of a Consolidated 21–103 mass spectrometer to determine trace impurities in oxygen. He allowed about 200 ml. of gas at 80 mm pressure (i.e. about 20 c.c-atmospheres) to react with sodium-potassium alloy, reducing the oxygen partial pressure to less than 0.01 micron. The remaining, unchanged gas was expanded into the mass spectrometer reservoir for analysis in the usual way. Typical results were: N_2, 5.5 p.p.m.; Ar, 0.7 p.p.m.; Kr, 8.6 p.p.m.; Xe, 1.1 p.p.m.

In the vacuum industry

The release of gases in a television tube will reduce its working life and it is important that when new materials are introduced, as in colour picture tubes, adequate information exists about the residual gases. Moscony and Turnbull[32] described how they built a small mass spectrometer on to the side of a picture tube and studied the effect of barium "gettering".

Evans and Collins[33] have used 60°, 7.5 cm radius, sector mass spectrometers in studies on residual gases in vacuum equipment, investigating the outgassing of the ionization gauge grid and investigating the nature of the residual gases after evacuation by Penning and radial field pumps. Day and Evans[34] described a small 180° analyser, radius 1.5 cm, resolving power nearly 50, which can give accurate partial pressure indications and is useful as a leak detector with hydrogen or helium as probe gases. They mentioned the use of one of these mass spectrometers to monitor the oxygen and water vapour in a vacuum-coating plant where semi-conductor oxide resistance films were being deposited.

INDUSTRIAL PROCESS CONTROL

At the outset, I referred to the direct use of mass spectrometers on-line in industrial processes. Both Consolidated and AEI have produced process analysers (e.g. AEI's MS 6) but their lack of popularity may have been caused as much by sampling problems as by mass spectrometric problems. Schuy[35] has described the use of an M.A.T. GD 150/4 for the continuous

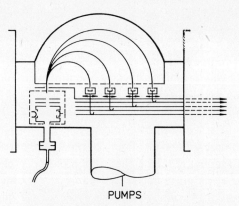

PUMPS

Figure 23   GD 150/4 mass spectrometer

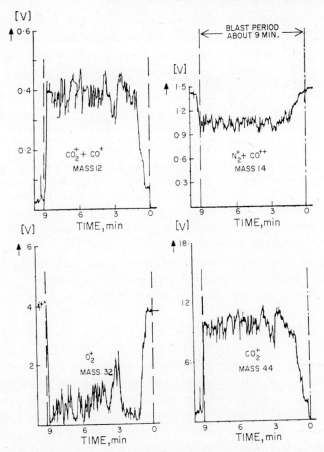

Figure 24 Continuous monitoring of waste gas (from reference 35)

Time-point	Gas composition, %				
	CO_2	CO	O_2	N_2	Σ
1	6.43	1.0	21.5	72.8	101.7
2	35.5	9.6	0.9	54.8	100.8
3	45.6	17.2	0.3	36.6	99.7
4	42.7	9.1	0.8	47.9	100.5
5	38.4	~0	10.8	50.8	100.0

Figure 25 Typical converter waste gas analysis during a blast period of about 9 minutes duration (from reference 35)

analysis of converter waste gas in steel making. The GD 150/4 (Figure 23) is a small 180° analyser mass spectrometer which can collect four ion beams simultaneously. Continuous monitoring (Figure 24) follows gas changes with the blasting process, producing typical analyses, as given in Figure 25. For further interesting examples, see the lecture by Mr. G. Ball to this Summer School.

APPLICATIONS IN THE METALLURGICAL, PIGMENT, CERAMIC AND SEMI-CONDUCTOR INDUSTRIES

I have included the above industries as examples of the use of spark-source mass spectrometers. I shall not go into the geometry or operation of these machines, on the assumption that others at this school, more qualified than I, will do so. Spark-source instruments are excellent for detecting very low concentrations of impurities in inorganic materials. The problems are to gain adequate reproducibility, because so little sample is consumed, to ensure the correct relative sensitivities and to render non-conductors conducting (solved by adding graphite, silver or gold).

Figure 26 shows the analysis of an aluminium sample[36] by spark-source mass spectrometry, and Figure 27 the analysis of an iron bar[37].

Spark-source mass spectrometry has the great advantages of speed and economy of effort over wet chemical analyses and for the determination of impurities in titanium dioxide pigments at the part-per-million level, Jackson and Whitehead[38] found the reproducibility of 15 % perfectly acceptable. Typical results are shown in Figure 28. By introducing an ion beam chopping technique[39], so that more sample was consumed for a given charge collected, they improved their reproducibility for the determination of calcium and phosphorus in titanium pigments from standard deviations of 12 to 13 % to 4 to 5 %.

Enamel frits are difficult to analyse because the mass spectrographic technique is really too sensitive* for components at the per cent level. Excessive dilutions with graphite or irreproducibly short exposures seemed indicated, until Jackson[40] found that the weaker doubly-charged ions gave results comparable in accuracy and reproducibility with those from longer wet

---

* Addink reported (*Inst. Phys. Conf.*, Exeter, July 1964) that mass spectrography was on average 150 times more sensitive than emission spectrography and over 5000 times more than X-ray fluorescence.

chemical methods. Typical results are shown (Figure 29) of the oxide concentrations found.

Honig, of RCA Laboratories, has given a useful review of the experimental problems of inorganic analysis by mass spectrometry[41]. In it he

Impurities in aluminium, p.p.m.[a]			
Cu	140	Cr	122
Ni	140	V	155
Mn	125	Ti	150
		Mg	99

[a] From reference 36

Figure 26

describes the analysis of a sample of doped silicon, prepared by Texas Instruments Inc., the semiconductor manufacturers. The results are shown in Figure 30. Further experiments on gallium arsenide showed that about half the Periodic Table could be detected at impurity levels of 1 to 10 p.p.b. (American billion $= 10^9$) and the remainder at the 10 to 100 p.p.b. level. Such sensitive analyses are invaluable in the production of reliable transistors and other semi-conductors.

B	—	Mn	1
Na	—	Co	5
Mg	0.2	Ni	4
P	0.2	Cu	1.1
S	0.35	Zn	—
Cl	0.8	Ga	<0.25
K	0.06	Ge	—
Ca	0.1	As	0.2
Ti	0.07	Nb	0.09
V	—	Mo	0.7
Cr	3	U	—

[a] From reference 37

Figure 27 Batelle iron bar: Spark source analysis, p.p.m.
(mass spectrometer: MS7)[a]

	p.p.m.	%		p.p.m.
B_2O_3	0.07		ZrO	120
MgO	40		Mo_2O_3	10
Al_2O_3		1.8	Sn	90
SiO_2		1.6	Sb_2O_3	550
P_2O_5		0.15	BaO	5
K_2O	80		LaO	0.21
CaO	550		CeO	0.4
V_2O_5	5		HfO_2	8
Cr_2O_3	2.2		Ta_2O_5	65
Mn_2O	0.14		W_2O_3	60
ZnO		0.94	Pb	95
As_2O_3	15		UO	0.8
SrO	3		ThO_2	0.8
			Fe	55

[a] From reference 38

Figure 28   Impurities in titanium dioxide pigments[a]

	% w		% w		% w
TiO_2	16.8	ZnO	4.5	P_2O_5	1.25
B_2O_3	15.1	K_2O	7.1	As_2O_3	0.9
Al_2O_3	0.29	Na_2O	9.0	MgO	1.05

Balance: SiO_2 (about 44% w)

[a] From reference 40

Figure 29   Analysis of vitreous enamel frits[a] (based on intensity of doubly-charged ions, to reduce need for excessive dilution with graphite)

B	3–6	Sb	200–600
Ga	1200	Al	20
As	30–60	S	100–300

[a] From reference 41

Figure 30   Analysis of doped silicon[a]
(atomic concentrations $\times$ 10)

CONCLUSIONS

I hope that in this survey of many of the industries using mass spectrometers (and I have omitted the whole field of nuclear energy) I have given some indication not only of the variety of problems tackled in so many diverse enterprises but also how the many different types of mass spectrometers are used. The problems facing the mass spectroscopist are so many and varied that he must use not only the most appropriate instrument for his work, but also bring to bear all the varied techniques and modes of operation of which his instrument is capable. This is even more the case where he is trying to use a mass spectrometer which is less than optimum in design for the task in hand. Many hours have been spent trying to squeeze a little more resolving power, a better signal-to-noise ratio, a little faster response from an instrument working on its design limit.

The most modern machines are so versatile, have such high resolution and, with on-line computing facilities, such a high rate of output that they make a significant contribution to the solution of industrial production and research problems.

References

1. H.J.Coleman, D.E.Hirsch, and J.E.Dooley, *Analyt. Chem.*, **41**, 800 (1969).
2., 3. B.F.Dudenbostel and P.J.Klaas, *Advances in Mass Spectrometry*, **1**, 232 (1959).
4. J.M.B.Bakker, A.Quayle, and L.V.Ashcroft, *XII Colloquium Spectroscopeum Internationale*, Exeter, p. 402 (1965).
5. K.Halbfast, *Advances in Mass Spectrometry*, **4**, 3 (1968).
6. A.L.Burlingame, *Advances in Mass Spectrometry*, **4**, 15 (1968).
7. A.Carrick, *Int. J. Mass Spec. Ion. Phys.*, **2**, 333 (1969).
8. F.W.McLafferty, R.Venkataraghavan, R.J.Klimowski, and J.E.Coutant, *ASTM E-14 Meeting*, Dallas 1969, Paper 16.
9. E.J.Gallegos, J.W.Green, L.P.Lindeman, R.Letourneau, and R.M.Teeter, *Analyt. Chem*, **39**, 1833 (1967).
10. G.L.Kearns, N.C.Maranowski, and G.F.Crable, *Analyt. Chem.*, **31**, 1646 (1959).
10a. L.R.Snyder, *Analyt. Chem.*, **41**, 1084 (1969).
11. *Report of Inst. Pet. Spectrometric Panel* ST.G4 to ASTM E-14 (1 December, 1965).
12. W.L.Mead, *Analyt. Chem.*, **40**, 743 (1968).
13. A.Fish, *Proc. Roy. Soc.*, **A298**, 204 (1967).
14. A.Fish, *Combustion and Flame*, **13**, 23 (1969).
15. J.Jorg, R.Houriet, and G.Spiteller, *Monats. Chem.*, **97** (4), 1064 (1966).
16. R.B.Delves and V.P.Williams, *Analyst*, **91**, 779 (1966).
17. J.N.Damico, R.P.Barron, and J.A.Sphon, *Int. J. Mass Spec. Ion. Phys.* **2**, 161 (1969).

17a. J. Robinson, A. Richardson, A. N. Crabtree, J. C. Coulson, and G. R. Potts, *Nature*, **214**, 1307 (1967).

17b. J. H. Koeman, M. G. Ten Noever De Brauw, and R. H. De Vos, *Nature*, **221**, 1126 (1969).

18. J. W. Cornforth, R. H. Cornforth, C. Donninger, and G. Popjak, *Proc. Roy. Soc.*, **B 163**, 492 (1965), and adjacent papers.

19. F. T. Addicott, K. Ohkuma, and O. E. Smith, *Tetrahedron Letters*, **29**, 2529 (1965).

20. J. W. Cornforth, B. V. Milborrow, G. Ryback, and P. F. Wareing, *Nature*, **205**, 1269 (1965); J. W. Cornforth, B. Milborrow, and G. Ryback, **206**, 715 (1965).

21. H. G. Langer, R. S. Gohlke, and D. H. Smith, *Analyt. Chem.*, **37**, 433 (1965).

22. R. S. Gohlke and H. G. Langer, *Analyt. Chem.*, **36** (10), 25A (1965).

23. J. Oro, J. Han, and A. Zlatkis, *Analyt. Chem.*, **39**, 27 (1967).

24. J. H. Beynon, R. A. Saunders, and A. E. Williams, *Mass Spectra of Organic Molecules*, (Elsevier, Amsterdam, 1968), p. 322.

25. *ibid.*, p. 342.

26. N. C. Rol, *Advances in Mass Spectrometry*, **4**, 215 (1968).

27. R. E. Deck and S. S. Chang, *Chem. and Ind.*, 1343 (1965).

28. W. H. McFadden and R. Teranishi, *Nature*, **200**, 329 (1963).

29. M. L. Aspinal, *Analyst*, **91**, 33 (1966).

30. R. T. Parkinson, *Talanta*, **14**, 1037 (1967).

31. J. Roboz, *Analyt. Chem.*, **39**, 175 (1967).

32. J. J. Moscony and J. C. Turnbull, *Nuovo Cimento, Suppl.*, **5**, 93 (1967).

33. S. Evans and R. D. Collins, "Mass Spectrometry", ed. R. Brymer and J. R. Penney, (Butterworths, London, 1968), p. 77.

34. M. Day and S. Evans, "Mass Spectrometry", ed. R. Brymer and J. R. Penney, (Butterworths, London, 1968), p. 117.

35. K. D. Schuy, *Z. Instru.*, **75**, 190 (1967).

36. N. W. H. Addink, *XII Colloquium Spectroscopeum Internationale*, Exeter, p. 651, 1965.

37. J. M. McCrea, *Appl. Spec.*, **23**, 55 (1969).

38. P. F. S. Jackson and J. Whitehead, *Analyst*, **91**, 418 (1966).

39. P. F. S. Jackson, J. Whitehead, and P. G. T. Vossen, *Analyt. Chem.*, **39**, 1737 (1967).

40. P. F. S. Jackson, *Conference of Brit. Ceramics Res. Assoc.*, October 1968.

41. R. E. Honig, *Annals New York Acad. Sci.*, **137**, 262 (1966).

Some considerations of the naive analysis of structure

R. I. REED and D. H. ROBERTSON

Department of Chemistry, University of Glasgow
Glasgow W2, Scotland, U.K.

It is always easier to explain the presence of a particular peak in a mass spectrum than to predict the main peaks to be expected in the spectrum of a particular compound and it is only when this latter problem is attempted that the naivety of our present level of understanding is fully realised.

JOHN BEYNON

The use of mathematical analysis is well established in the fields of communication theory and data reduction; computer programmes are available for handling this kind of information and the expertise whereby the technique may be applied to practical problems has reached a highly developed state. The authors of this presentation have been concerned with the possibility of applying one or more of these mathematical methods to the analysis of mass spectra. Using elementary set theory as a starting point, a straightforward method is developed whereby mass spectral data can be manipulated to provide diagnostic criteria for the various ions in the spectrum under consideration.

The problem of identification of an unknown by comparison with a data bank of known spectra has received considerable attention. The specific technique employed for the retrieval depends whether one is working with high or low resolution data. A promising beginning has been made with respect to low resolution spectra[1]; the situation with high resolution is considerably more difficult to define. Most recently the principle of artificial intelligence has been applied to these analyses[2]. In the latter case the computer is provided with a programme which allows it to estimate the agreement of the unknown spectrum when it is compared against a so-called training set of spectra; the computer is thus able to decide to which classification the unknown belongs.

Inasmuch as modern mathematical techniques are highly dependent upon the use of computers for the purpose of reducing and processing gross quantities of otherwise unwieldy data, it has been considered desirable to develop a method of mass spectral analysis which would incorporate the binary principle. Appropriate use of set theory allows this to be realised. Some of the investigations, especially those involving cycloalkanes, suggest that strict adherence to a binary classification may not be feasible.

In its simplest form, naive analysis allows one to consider the ions produced in a mass spectrometer as subsets or (in the case of an individual ion) as a member of the universal set of all possible ions which may arise from the fragmentation of an organic molecule. With this basic premise we have used naive set theory successively to examine alkanes, alkenes and cycloalkanes. As the work has progressed it has been found that the simple criterion which was assumed for alkanes does not completely satisfy the latter two classifications of hydrocarbons. The problem of a criterion or criteria upon which a mass spectral analysis is based has not been solved with complete satisfaction; however, the development of these applications will be presented as they have occurred in the laboratory. How they may be used at present in spectrum analysis will be indicated and areas for future investigation will be suggested, whereby application of the technique may be extended to further classes of organic compounds.

The source of data in each case is the A.P.I.[3] file of mass spectra, the use of which assumes the comparison of an unknown with one or more reference compounds. The comparison is made ion for ion between unknown and reference compound. In the ideal case, the structure of the unknown may be deduced by determining whether or not it has a positive correlation with an ion in the reference compound. (It is implicitly assumed that each ion which appears in the spectrum of the reference compound has a particular relationship to the structure of the molecule from which it comes.) There is variable success in "predicting" an unknown structure depending on which reference compound is used and what criterion (criteria) is (are) employed. The reference compound which has been selected for the initial analyses is the straight-chain alkane containing the same number of carbons as the "unknown" structure with which it is compared. Although some experimentation has been made with the use of different compounds, the results are far from definitive. A somewhat more detailed discussion of this problem is treated in the section on alkenes.

Inasmuch as we can freely define what constitutes the membership of a set, the potential inherent in the use of naive set analysis is great. The universal set may be looked upon as containing all the possible ions resulting from fragmentation in the mass spectrometer of all the known organic compounds (we tacitly assume that the treatment is restricted to the realm of organic chemistry). In the simplest situation which can be imagined, one could select a subset from the universal set which would embody all the ions necessary to provide a unique diagnostic for a given class of compound i.e. set A is composed of all the ions which are required to identify unambiguously the class, hydrocarbons. The set A will likewise be subject to subdivision into further subsets, which represent, in the case of the above example, such classifications as alkanes, alkenes, alkynes, cycloalkanes and cycloalkenes.

There are some pitfalls which one may fall into when making the translation from naive set theory to more rigorous considerations. Among these are the axiom of choice, Zorn's lemma and the well-ordering theorem. However, at the present point of development there is need for little more than an awareness of these pitfalls. As the set theory approach is formalised in future work mere awareness will not be enough.

ALKANES

Before the actual manipulation of data is begun, it is necessary to provide a series of assumptions which in this initial case are applicable to alkanes in particular. Later in this article we shall discuss their application to the alkenes and cycloalkanes.

These initial assumptions (relative to alkanes) are:

1. fragmentation of a parent ion is favoured at points of chain branching with preferred elimination of the largest branch,

2. ions of even mass often arise in association with such fissions, due to concomitant hydrogen rearrangements, and

3. the parent molecular ion of a branched chain alkane is less abundant than that of the isomeric straight-chained alkane.

It is also necessary to consider the fact that isomeric ions may be formed, in which case the observed ion currents (as obtained in these calculations by the addition of the individual normalised ion intensity values taken directly from the A.P.I. data sheets) for any given mass will be represented by the sum of the individual isomeric ions.

In so for as the use of set theory implies a collection of all possible ions which can be conceived as being formed from a given structure, regardless of whether or not they are actually produced in the mass spectrometer, a criterion must be adopted for rejecting unacceptable ions i.e. those which do not actually appear in the list of m/q^* ratios in the A.P.I. tables of spectra. In addition there is also the possibility of sequential reactions occurring; however, it is obviously not possible to make any allowances for this phenomenon without first assuming some information about the structure which one is attempting to determine. In the case of an actual analysis, structural information about the unknown would of course be absent. As the rationale behind the technique is to provide a means of comparison between reference compound and unknown which will allow identification of the latter, such a possibility shall not be considered in this treatment. Thus it is that the ions which result in a given spectrum are used for analysis without scrupulous consideration of their origin.

The approach to saturated hydrocarbons (alkanes) is based on the observation that the major differences which occur among their respective spectra will be found in the branched chain compound of a given carbon number. It is in this manner that a means of locating the branching point in an isomeric alkane is obtained i.e. by comparing its principal ions of the general formula $C_nH_{2n+1}^{+\cdot}$ with the comparable ions in the straight chain compound.

Based on the assumption that total ion current is the same for each isomer, intensities of ions formed by fragmentation at a branch will be greater than the same empirical ion from the straight chain reference standard. Analogous behaviour has been observed with regard to the alkenes.

A binary oriented ratio is employed which works out to be "1" if the ion has a diagnostic value relative to the reference compound and "0" if it does not. In terms of our set theory treatment, it is observed that ratios with a value of "1" are found to have a symmetrical distribution around the centre of the molecule; this centre (again in terms of set theory and for the $n-1$ ion, where n = number of carbons) is at $n/2$ for ions from odd numbered

---

* The use of m/q follows the suggestion of the American Institute of Physics.

carbon compounds. It should be noted in regard to the f values, which are subsequently developed, that their distribution is likewise symmetrical around the centre of the molecule. (Please see note in Table 1.)

The ionization process for the molecule XY is considered as:

$$XY + e \rightarrow X^+ + \dot{Y} + 2e \quad \text{or} \quad \dot{X} + Y^+ + 2e$$

Table 1

C_nH_{2n+1}	l	l_E	l_K	r_l	f_l
15	1	3.90	3.00	1.08	1
29	2	42.50	34.50	1.02	1
43	3	100.00	100.00	0.83	0
57	4	67.30	34.20	1.64	1
71	5	3.05	28.30	0.09	0
85	6	48.60	29.50	1.37	1
99	7	0.76	0.07	9.03	1
114	(8)	2.99	6.74	0.37	(1)
Σ	.	485.26	403.71	.	.

$$r_l = \frac{l_E}{l_K} \cdot \frac{\Sigma_K}{\Sigma_E} = \frac{l_E}{l_K} \cdot \frac{403.71}{485.26}$$

K = the known reference standard; in this case it is n-octane.

E = the unknown compound whose structure is sought; here, it is 3-methylheptane.

Σ = Total ion current taken as the sum of normalised individual ion intensities.

l_E = an ion intensity (unknown).

l_K = an ion intensity (reference compound).

Values for f_l which are diagnostic for the ions are distributed symmetrically around the value for C_4, which represents both C_4 fragments arising from bond cleavage between C_4 and C_5 (n-octane) and/or C_3–C_4 (3-methylheptane).

It is assumed that bond dissociation energies of the carbon–carbon bonds in neutral alkane molecules occur in decreasing order: primary > secondary > tertiary and thus the concentration of X and/or Y radicals would be greater for the branched chain structure, assuming the concentration of straight- and branched chain species to have been the same before ionization. Abundance of the branched-chain ion is represented by $[X_E^+]$, that of the straight chain reference compound by $[X_K^+]$. Based on bond energies, we may say that the ratio of branched- to straight-chained structures is equal to or greater than "1", i.e. $X_E^+/X_K^+ \geq 1$. The same type of expression may be written for the "Y" ion, i.e. $Y_E^+/Y_K^+ \geq 1$. Thus, in general, ion abundance ratios which are greater than unity will appear in pairs, and the sum of their masses will of course equal the mass of the parent molecular ion (see Table 1).

For the purposes of this investigation it shall be assumed that ionization cross section depends to a greater extent upon constitution of a molecule than upon its structure and that total ion current represents a reasonable approximation of the ion cross section. Ötvos and Stevenson[4] suggest that there is simply a proportionality between cross section and total ion current. It will be pointed out in the discussion of cycloalkane analysis that for the purpose of obtaining diagnostic values, the original approximation is satisfactory.

Ideally, the ion cross section (σ) and total ion current (Σ) for the isomeric hydrocarbons would be the same. For the purpose of the argument the observed values are treated as if they were mathematically exact, although in actual fact σ is only approximately so for different isomers. Therefore, in the case where differences are observed, the ion currents of the branched-chain alkanes are sealed down to bring them into line with those of the *n*-alkanes; in other words, the branched chain molecules are normalized to the straight chain molecule having the same number of carbons.

For a general analysis, the series of alkyl ions and their associated abundances have been considered the most useful. Not only can they be assigned in pairs by the relationship $P = X^+ + Y^+$ but they are usually the most important of all the ions associated with a given carbon number.

The following simple notation is hereby adopted which underlies the connexion between the ion formed from the parent:

$$C_nH_{2n+2}^{+\cdot} = P^{+\cdot} = (C_lH_{2l+1} + C_mH_{2m+1})^{+\cdot} = l + m$$

The shorthand nomenclature for normalizing to the straight chain compound is as follows:

$$\left(\frac{l_E}{l_K}\right)\frac{\Sigma_K}{\Sigma_E} = \frac{[C_lH_{2l+1}^+ \text{ (branched)}]}{\Sigma \text{ (branched)}} \times \frac{\Sigma \text{ (straight chain)}}{[C_lH_{2l+1}^+ \text{ (straight chain)}]} = r_l$$

Finally, we adopt one additional index f_l which is assigned the value of "1" ir r_l is equal to or greater than "1" and a value of "0" if r_l is less than "1".

A condensed form of the nomenclature, using 3-methylheptane as an example is given in Table 1. Again, only the "*l*" ion is represented. The reference compound, denoted by subscript K, is *n*-octane.

Distribution of the f_l values is now further developed. If ionic species of the general form $C_nH_{2n+1}^{+\cdot}$ are arranged equidistantly along an axis, they should group in pairs e.g.

$$l = 1 \quad \text{and} \quad l = n - 1$$
$$l = 2 \quad \text{and} \quad l = n - 2$$

based on the fact that a compound of carbon number n produces an ion of carbon number l and an ion of carbon number $n - l$. The values of l which are equidistant from the centre of the molecule should have the same f_l values. In the case of 3-methylheptane,

$$l = 2, \quad f_2 = 1$$
$$l = 6, \quad f_6 = 1$$

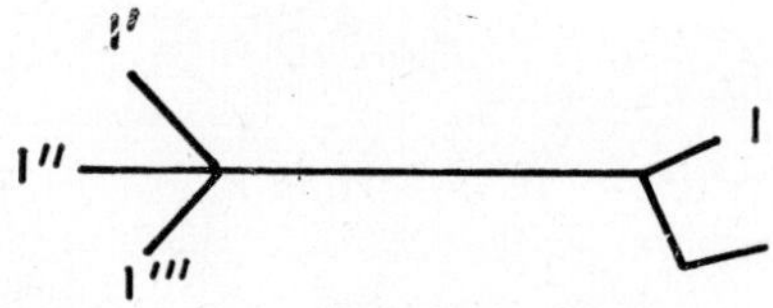

Figure 1

both of which are equidistant from the centre of the molecule which is at carbon 4. (See note in Figure 1.) In the determination of the structure of the molecule, one derives $l + m$, where $1 < l$, $m < n/2$ or $(n - 1)/2$ and $m = n - l$. Again referring to 3-methylheptane, there are three such pairs:

$$4,4; \ 2,6 \ \text{and} \ 1,7$$

The use of set theory also allows one to generate the possible structure of ion groupings from an unknown. These groups are of the form $\cdot CH_3$, $\cdot C_2H_5$, etc. and are represented by the arabic numerals *1, 2* etc., corresponding to the carbon number. If a radical appears more than once in a single molecule, it must be given a separate notation each time it appears. For example, in 2,2' 3-trimethylpentane, there are four different methyl groups (as indicated in Figure 1) which may be uniquely identified as 1, 1', 1", 1'''. This collection of symbols represents subsets of the complete or universal set which is itself represented by P or the appropriate carbon number; this number would of course be *8* in the case of the octanes. In addition, a null or empty set which contains no elements must be considered.

Each symbol (1, 1′, 1″, 1‴, 2, 3, etc.) represents a proper subset of the universal set; there are also two improper subsets, the null set and the universal set itself. With the exception of the null-set, each set or subset contains elements which are identified with the number of carbons contained in the radical under consideration. Thus, trimethylpentane is represented by the universal set *8* as well as by at least one subset *4*; these sets contain *8* and *4* elements respectively.

$$A \cup B \cup \cdots \cup N = \sum_{A}^{N} A - \sum_{2}^{n} C_{2} (A \cap B)$$

$$+ \sum_{ABC}^{LMN} {}^{n}C_{3} (A \cap B \cap C \cdots$$

$$(-1)^{n-1} [A \cap B \cdots \cap N]$$

Figure 2

In a system of finite order, an important relationship exists between sets and their subsets. Capital letters represent sets or subsets, $\cup$ represents the union of sets and $\cap$ represents the intersection of sets. Again referring to 2,2′,3-trimethylpentane in which the methyl group occurs four times, we depict in Figure 3 a set representation of the compound. Automatically, 4 distinct heptyl groups can now be generated; these are represented by 7, 7′, 7″ and 7‴. (Since there are four distinct methyl groups, there will be an equal number of heptyl groups i.e. groups of the form $P-CH_3$, where P denotes the parent. In Figure 3, the solid line represents the portion of the structure which is common to each of the heptyl groups; i.e., a butyl chain. The numeral *7* represents the total structure less the methyl indicated by the dotted line; 7′, the total structure less the methyl group indicated by the dashed line, etc.

In order to make further analysis easier, we introduce the concept of set intersection. Since it is clear that none of the subsets 1, 1′, 1″, 1‴ can inter-

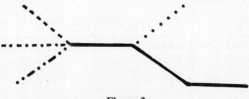

Figure 3

sect with each other, we may say that the intersection of 1 with 1′, for instance, is the null set. However, this condition does not apply to members of the subset represented by the numeral 7 because the intersection of 7 with 7′ will include that part of the molecule which is common to both.

Consideration of the methyl radicals can be simplified by considering the property of sets which states that the intersection of any set with the null set is itself null. It is clear that the subsets 1, 1′, 1″ and 1‴ cannot intersect one with the other, so that $1 : 1′ = \varnothing$ for all intersections of methyl groups;

$$\sum_{1}^{1‴} {}^{n}C_2 (1 \cap 1′) = \varnothing.$$

The situation with the heptyl (C7) radicals is different; taken three at a time, the intersection of 7, 7′ and 7″ yields a subset with 5 carbon atoms, as seen in Figure 4. An illustration of this application follows:

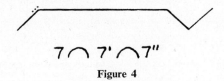

$$7 \cap 7′ \cap 7″$$

Figure 4

Consider a set A which is associated with a family of subsets $b - 1, b - 2$, etc., having the following properties:

1. A is a union of sets, for instance, b and c.

2. For any given pair of subsets b and c, either b equals c or $b \cap c$ equals $\varnothing$. Such a family of sets is called a partition of A. In octane, for instance, the subsets 4 and 4′ are the partition of P since $4 \cap 4′$ equals $\varnothing$.

Inasmuch as subsets 1 and 2 were observed in the actual analysis, we may write:

$$1 \cup 2 \cup 4 = 1 + 2 + 4 - \sum_{1,2}^{2,4} {}^{n}C_2 (1{:}2)\ 1{:}2{:}4$$

$$= 7 - \varnothing - 1{:}4 - 2{:}4 + \varnothing$$

Since $\qquad\qquad 1{:}2 = \varnothing \quad 4 = 7 - 1{:}4 - 2{:}4$

or $\qquad\qquad\qquad 3 = 1{:}4\ 2{:}4$

Figure 5

The intersection of 1 with 4 contains one element only, therefore $2 \cap 4$ contains two elements and the structure for the butyl or C_4 group emerges. Other possible subsets may be deduced in a similar way.

ALKENES

When olefines are considered, two features of the molecule become important from the standpoint of naive analysis: the points of chain branching and the location of the double bond. The following rules may be used with both alkanes and/or alkenes:

1. the parent molecular ion is biggest for the straight chain isomer
2. loss of a C_1 fragment is likely only for a CH_3 side chain
3. fragmentation is in general most likely at highly branched carbons
4. C_3 and C_4 fragments are always prominent in alkane spectra.

Owing principally to the presence of the unsaturation, ions from an alkene are more stable than and occur with greater relative intensities than analogous ions from alkane structures. It is also observed that the abundance of the parent peak in an alkane, relative to its fragment peaks, is greater than in the case of saturated hydrocarbons. It is unfortunate for the purposes of establishing rules for alkenes, that the latter trend is found to reverse itself above C_6; this may be related to the observation that low mass unsaturated ions are more stable than the saturated ions of comparable mass (i.e. comparable ions from paraffins). Generally speaking, one may say that the preferred point of fragmentation in the olefine molecule is allylic to the double bond. In addition there exists a greater tendency for rearrangement to occur in the mass spectrometric analysis of unsaturated hydrocarbons with the net result that their mass spectra are less sensitive to variations in molecular structure. Given these considerations, a series of olefine spectra will be studied (olefines which are structually similar) with attention to the 41 and 55 peaks and the two series:

a)   41, 55, 69, 83, etc.

b)   42, 56, 70, 84, etc.

The series of ions of the form C_nH_{2n-1} (a, above) predominates but with a decrease in magnitude with increasing molecular weight; there is at the same time an increase in prominence of the molecular ion.

Table 2

Compound (A.P.I. no.)	PAIR	a	b
99	1,5	+, +	−, +
	2,4	+, +	−, +
	3,3	0.73, 0.73	0.53, 0.53
100	1,5	+, +	+, +
	2,4	+, +	+, +
	3,3	+, +	+, +
101	1,5	+, +	+, +
	2,4	+, +	+, +
	3,3	+, +	+, +
102	1,5	+, +	+, +
	2,4	0.86, +	0.77, −
	3,3	+, +	+, +
103	1,5	+, +	+, +
	2,4	−, +	−, −
	3,3	+, +	+, +
104	1,5	+, +	+, +
	2,4	0.88, +	0.78, +
	3,3	+, +	+, +
276	1,5	+, +	+, +
	2,4	0.85, +	0.77, +
	3,3	+, +	+, +
278	1,5	+, +	0.84, +
	2,4	−, +	−, −
	3,3	+, +	+, +
279	1,5	+, +	+, +
	2,4	−, +	−, −
	3,3	+, +	+, +
399	1,5	+, +	+, +
	2,4	0.99, +	0.88, +
	3,3	+, +	+, +
524	1,5	+, +	+, +
	2,4	−, +	−, −
	3,3	+, +	+, +

Table 2 (cont.)

Compound (A.P.I. no.)	PAIR	a	b
525	1,5	+, +	+, +
	2,4	−, +	−, −
	3,3	+, +	+, +

99 – trans-hex-2-ene
100 – trans-hex-3-ene
101 – 3-methylpent-1-ene
102 – 4-methylpent-1-ene
103 – 2-methylpent-2-ene
104 – 3-methyl-cis-pent-2-ene
276 – 2-methylpent-1-ene
278 – 4-methyl-trans-pent-2-ene
279 – 4-methyl-cis-pent-2-ene
399 – cis-hex-3-ene
524 – 2,3-dimethylbut-1-ene
525 – 2,3-dimethylbut-2-ene

Column *a* represents data for n-hexane; column *b* data for hex-1-ene
+ indicates that a correlation exists; − that a correlation is absent; where decimal fractions are given, the intention is to suggest that values of the ratios which are less than 1 (see the "1" or "0" criterion in text) may have diagnostic value, especially when they are near 1.00 in value

It must be noted that the even mass peaks may be confused with those formed by double fragmentation in paraffins unless a metastable transition is observed. The hydrogen rearrangements which are so very prevalent in alkenes result in migration of radical sites along the chain; therefore the location of the double bond has very little influence upon the spectrum except in those instances where the double bond has undergone tetrasubstitution.

Another basic problem in the use of the naive analysis of spectra lies in the selection of the reference compound (see treatment of alkanes) with which an unknown is compared for purposes of getting diagnostic ions. Thus it is that the question again arises: How close in structure to the unknown compound must the reference compound be in order to provide such ions as occur in the observed spectra?

As a means of comparing reference compounds, the diagnostic pairs resulting from a C_6 hydrocarbon (see alkane data) are listed in Table 2, namely: 1,5; 2,4 and 3,3. Each C_6 monoalkene has been compared for these pairs using (a) *n*-hexane and (b) hex-1-ene as reference compounds. Based on our earlier comments as regards the general lack of influence which saturation has in alkene spectra, it would not be supposed that any significant difference in diagnostic value of the various C_nH_{2n-1} ions would manifest itself when the reference compound is changed. The data in Table 2 indicate that such a difference does exist (specifically with respect to the 2,4 pair of ions) depending upon which of the two reference compounds is used. The 2,4 pair corresponds to ions at masses 27 and 55 respectively, both of which would be expected in alkene spectra since they are of the form C_nH_{2n-1}.

Study of the twelve C_6 alkenes in the table indicates that if the molecule is symmetrical with respect to the distribution of side chains or is straight chained (consider compounds 99, 101, 104, and 399) the same diagnostic value is obtained for the 2,4 ion pair from both reference compounds. However, in the case of unsymmetrical molecules (see compounds 102, 103, 278, 279, 524 and 525) one fails to obtain diagnostic value for the 2,4 ion pair when hex-1-ene is used as the reference compound; this failure occurs specifically with the C_4 member of the pair. One is thus led to postulate that in the treatment by naive analysis the unsaturated function assumes a degree of importance in establishing the diagnostic value of certain ions. It is further observed that those compounds with a low intensity in their mass spectra for the C_4 ion at mass 55 are the same compounds which fail to give diagnostic value for the C_4 ion when the reference compound is changed from *n*-hexane to hex-1-ene.

In attempting to apply set theory to analysis of mass spectra, there is an awareness of the shortcomings in general applicability of this concept to all types of compounds. It was felt initially that hydrocarbons offered the simplest group of spectra with which to test the theory. Particularly after the observation has been made that the olefine spectra, when subjected to set theory analysis, are sensitive to structural variation, it seems possible that the simplicity of structural elements (C and H) conceals much more than was expected from the early investigations concerning alkanes.

CYCLOALKANES

In the preceding treatment of alkanes and alkenes, the basis of analysis was the occurrence of preferred fragmentation at a point of chain branching, to yield ions of the formula $C_nH_{2n+1}^+$ and $C_nH_{2n-1}^+$ respectively. Use of these ions has been shown to be satisfactory for generation of a structural formula, at least under those conditions imposed upon the analysis in this early stage of development.

When dealing with alkanes, chain branching constituted the only possible difference in structure between the reference straight chain hydrocarbon and the unknown alkane being considered. In the case of the alkenes, position and extent of unsaturation becomes a second factor in comparison with the reference compound and hence the need for more exhaustive investigation of the effect upon analysis of a change in reference compound.

As only the simplest case of an alkyl substituted fully saturated ring system is studied (in the initial approach to cyclic hydrocarbons) the types of ions which must be considered for an analysis are still only two: a) the alkyl ions which arise from the alkyl chains in the original cyclic structure and b) an alkenyl series, resulting from any sequence containing an alicyclic ring. In the earlier work, if r_l was greater than or equal to unity $f = 1$, otherwise $f = 0$.

In the present series, (inasmuch as the reference compound of choice is still the straight chain alkane of the same carbon number as the cyclic-molecule being analysed) one is comparing the abundance of the corresponding alkenyl ions from a cyclo- and a normal alkane.

The alkenyl ion from a cycloalkane is formed by simple fission; however, the distribution of positive charge between the resultant moieties is not equal. Other considerations being the same, the cyclic portion of the molecule will yield the preponderance of the fragment ions, if only because of its lower ionization potential. Formation of an alkenyl ion from the reference alkane should involve loss of two hydrogen atoms from the alkyl ion, which is assumed to be formed first and thus the process does not lend itself to straightforward thermochemical arguments.

Even if both ions have the same ionic formula, it cannot be assumed "a priori" that there is any value in the ratio of their abundances.

Thus it is that we alter our test for diagnostic requirement of a ratio value for any given ionic species; the "0" or "1" criterion is abandoned and the occurrence of a local maximum in the ratio is taken as the new criterion. However, the general conclusion that I_E increases at a branch point is unimpaired so that in a survey of ratios a local maximum may be considered in the same manner i.e. a local maximum rather than the "0" versus "1" evaluation is observed to reveal a branching point.

Ionization cross section values of Ötvos and Stevenson[4] were used in the calculations with alkanes and alkenes, in which cases they were assumed to approximate the values for total ion current obtained by addition of the intensity values for the ions appearing in the A.P.I. spectra. The availability of a distinct set of ionization cross section values (cf. Mann[5]) prompted a comparison of results using both sets. This comparison showed a difference of the order of 3%; but, as the positions of the maxima were not observed to change when Mann's values were substituted; investigations were continued with those of Ötvos and Stevenson.

The present method of calculation is thus as follows. Separate the ions of the mass spectrum into their various series $C_nH_{2n+1}^+$, $C_nH_{2n}^+$, etc. Let l_E represent the observed ion current for $C_nH_{2n+1}^+$ in the unknown and l_K that of the corresponding ion in the known; Σ_E is the total ion current of E and Σ_K the same value for K. The ionization cross-section for E is represented by σ_E and similarly for K. Then the corrected ratio values of the ion currents for the ion $C_nH_{2n}^+$ are $l_E \cdot (\sigma_E/\Sigma_E)$ and $l_K \cdot (\sigma_K/\Sigma_K)$. The comparison of these ion currents is more difficult than formerly, as both σ and Σ differ for the two compounds. To obtain a reasonable assessment, we must adjust the value of one (say E) to allow for these differences. Let $l_E \cdot \sigma_E/\Sigma_E = h$ and correct it to bring it to the same conditions as obtain for K. In correcting for the differences of ionization cross section we replace h by $h \cdot (\sigma_E/\sigma_K)$; by making a further correction for the total ion current, the expression becomes

$$h \cdot \frac{\sigma_E}{\sigma_K} \cdot \frac{\Sigma_K}{\Sigma_E}$$

The corrected value of the ion current for E becomes

$$l_E \cdot \frac{\sigma_E}{\Sigma_E} \cdot \frac{\sigma_E}{\sigma_K} \cdot \frac{\Sigma_K}{\Sigma_E}$$

and that for the ratio between the unknown and known ions

$$l_E \cdot \frac{\sigma_E \cdot \sigma_E \cdot \Sigma_K}{\sigma_K \cdot \Sigma_E \cdot \Sigma_E} \bigg/ \frac{l_K \cdot \sigma_K}{\Sigma_K}$$

Hence, by regrouping

$$\left(\frac{\sigma_E}{\sigma_K \Sigma_E \Sigma_E}\right) \cdot l_{E'} \cdot \sigma_E \Sigma_K / l_{K'} \sigma_K \left(\frac{1}{\Sigma_K}\right)$$

and rearranging

$$\left(\frac{\sigma_E \cdot \Sigma_K}{\Sigma_{E'} \cdot \sigma_K}\right) l_{E'} \cdot \sigma_E \Sigma_K / l^{K'} \sigma_K \Sigma_E$$

Accordingly, as was assumed in the previous communication[6], it is supposed that

$$\frac{\sigma_E \cdot \Sigma_K}{\Sigma_E \cdot \sigma_K} \approx 1$$

and that the wanderings from this simple theory are due to variations in ion-collection efficiencies and similar small perturbations whose origins are not fully understood.

It has been assumed that the error introduced by assuming a strict proportionality of the total ion currents of C_nH_{2n} and C_nH_{2n+2} with their ionization cross sections is small. Within these limits analyses have been carried out upon a series of cycloalkanes of the following general types: cyclohexyleicosanes, cyclic systems with a terminal ring, cyclic systems with multiple substitution and compounds having more than one cyclic group.

For the purposes of illustrating the method of data manipulation for this category of compound, the case of cyclohexyleicosanes is discussed. They are very appropriate because a series of position isomers of such compounds is available. The calculation is carried out in two stages. The first comparison is made between the alkyl ions of unknown and reference for the purposes of locating any branchpoints in the alkyl chains. Secondly, a comparable investigation is carried out for the alkenyl ions. The discussion of deductions for the isomer 5-cyclohexyleicosane will serve as an example.

The mass spectrum yields the molecular formula, $C_{26}H_{52}$. A series of abundant alkyl ions are observed at $m/q = 281$ (the most abundant) and corresponding to formula $C_{20}^+H_{41}$; $m/q = 155$ ($C_{11}^+H_{23}$); $m/q = 211$ ($C_{15}^+H_{31}$) and $m/q = 309$ ($C_{22}^+H_{45}$). The alkenyl ion sequence shows the presence of $C_6^+H_{11}$, $C_{11}^+H_{21}$, $C_{17}^+H_{33}$, and $C_{20}^+H_{39}$ ($m/q = 83, 153, 237,$ and 279). Excepting $C_{20}^+H_{39}$ and $C_{22}^+H_{45}$ these ions form a self-consistent elemental set which, by the methods used in the treatment of alkanes, yields a structure of the type $C_4H_9-CH(C_6H_{11})-C_{15}H_{30}$.

Three points must now be taken into consideration: 1) the nature of the C_6H_{11} ring system; 2) the possibility of chain branching in the $C_{20}H_{41}$ alkyl group and 3) the origin of the two ions $C_{20}H_{39}$ and $C_{22}H_{45}$. In reference to point 1), the cyclic group may be either cyclohexyl, an isomer of the methyl-

Table 3

l	*n*-Hexa-cosane.	Position of substitution (for cyclohexyleicosanes)					
		(-2)	(-3)	(-4)	(-5)	(-7)	(-9)
A.P.I. No.	669	1263	1264	1265	1487	1036	1037
Σ	578.02	846.69	937.37	876.21	1084.49	1318.06	1372.64
Alkyl ions							
3	14.800	0.399	0.407	0.451	0.351	0.303	0.277
4	17.300	0.244	0.271	0.307	0.241	0.247	0.248
5	9.913	0.197	0.217	0.258	0.240	0.214	0.215
6	6.834	0.180	0.207	0.248	0.179	0.215	0.211
7	1.510	0.170	0.219	0.259	0.218	0.319	0.298
8	0.986	0.176	0.214	0.264	0.211	0.303	0.328
9	0.723	0.165	0.216	0.258	0.209	0.302	0.316
10	0.562	0.215	0.201	0.262	0.199	0.272	0.280
11	0.441	0.146	0.200	0.256	0.212	0.222	0.287
12	0.369	0.139	0.195	0.239	0.187	0.288	0.279
13	0.318	0.135	0.183	0.231	0.187	0.278	0.235
14	0.282	0.130	0.170	0.233	0.176	0.252	0.216
15	0.246	0.121	0.165	0.204	0.189	0.190	0.211
16	0.220	0.107	0.142	0.203	0.159	0.139	0.134
17	0.201	0.079	0.140	0.156	0.091	0.077	0.085
18	0.185	0.093	0.238	0.083	0.058	0.048	0.074
19	0.173	0.020	0.040	0.035	0.025	0.015	0.030
20	0.159	**2.126**	2.076	**3.175**	**3.476**	**1.913**	**2.129**
21	0.144	0.024		0.028	0.015	0.006	0.003
22	0.125	0.017	0.300	0.016	0.138	0.004	0.003
23	0.083			0.266	0.013	0.011	
24	0.050		0.487	0.027			
25	0.005	0.694					
26	0.005	1.667	0.375	0.900		0.177	0.256
Alkenyl ions							
3	6.055	0.820	0.820	0.903	0.530	0.531	0.499
4	5.675	1.129	1.099	1.193	0.877	0.766	0.744
5	2.855	1.244	1.155	0.989	0.790	0.609	0.551
6	2.059	**4.480**	**3.232**	**3.317**	**2.199**	1.636	1.509
7	1.251	1.372	1.173	1.650	1.492	1.417	1.331

Table 3 (cont.)

	n-Hexacosane	Position of substitution (for cyclohexyleicosanes)					
		(−2)	(−3)	(−4)	(−5)	(−7)	(−9)
A.P.I. No.	669	1263	1264	1265	1487	1036	1037
Σ	578.02	846.69	937.37	876.21	1084.49	1318.06	1372.64
				Alkenyl ions			
8	0.512	**7.448**	1.464	1.405	1.575	1.790	1.656
9	0.213	1.175	**9.156**	1.787	1.338	2.389	2.276
10	0.078	1.330	2.281	**19.084**	2.021	2.681	3.455
11	0.042	1.530	2.496	3.291	**28.085**	3.161	4.122
12	0.028	1.753	2.717	3.103	3.083	3.819	3.881
13	0.017	2.073	0.353	3.804	3.238	**32.298**	4.361
14	0.012	2.241	3.646	4.167	3.243	5.556	5.125
15	0.007	2.871	5.268	5.429	5.154	6.139	**58.659**
16	0.005	3.472	6.000	7.067	5.699	5.398	5.897
17	0.003	4.419	9.167	10.000	6.964	5.294	8.286
18	0.003	3.954	7.917	7.111	5.536	3.971	97.429
19	0.002	2.069	6.563	6.338	4.324	5.556	5.106
20	0.002	4.138	8.750	5.667	8.378	116.444	2.553

$\sigma_E = 160.16$ in all instances, for the compounds are isomers.
$\sigma_K = 162.16$ for *n*-hexacosane, the reference compound.

$$\frac{x\text{-cyclohexyleicosane}}{n\text{-hexacosane}} = \frac{l_x}{l_{669}} \times \frac{\sigma_E}{\sigma_K} \times \frac{\Sigma_K}{\Sigma_E}$$

where x = position of substitution
l_x has two values, one for the alkyl ion series, one for the alkenyl ion series.
Ratios in bold type are considered the most reliable.
Values for mass 26 relate to the parent molecular ion and as such are not members of the alkyl ion sequences.

cyclopentyl group or (less likely) a substituted cyclobutyl group. The latter is unlikely on the basis of observed fragmentation patterns from cyclobutyl-containing systems. The present form of naive analysis is not able to distinguish reliably between the first two of these possibilities.

The second point, namely the possibility of chain branching, is considered to be remote because any such occurrences would imply a terminal cyclic group. If one assumes this fragmentation to provide all the necessary alkyl ions it would require two cleavages in the parent molecular ion, which phenomenon is excluded by our analytical principles.

In regard to the third point, the origins of $C_{20}^+H_{39}$ and $C_{22}^+H_{45}$ must be considered. The latter may readily be assumed to arise from cleavage within the ring, which would yield the observed $R = C_{20}H_{41}$. The complementary ion does not belong to either ion sequence here studied and thus is an example of the utility of the present status of naive analysis.

Despite the failure to provide complete explanation for all ions occurring in the course of a particular analysis, the method would appear to have advantages, especially its applicability in the technique of using a reference compound which may be analysed in admixture with the unknown. Techniques for subsequent recovery of the mass spectrum under these conditions are known. Thus, there is the possibility of overcoming problems due to short term instrumental variations and thereby allowing extension of the analysis to any mass spectrometer with the requisite minimum resolution.

A more detailed treatment of the cycloalkane systems has been published[6]. An initial effort, dealing with alkanes, has also appeared[7].

Worthwhile progress is currently being made with respect to a more complete treatment of systems with double bonds and those which include hetero-atoms i.e. atoms other than C and H. Although far from complete, these germinal ideas may serve as a beginning for a treatment of mass spectral data which will in time assume appreciable significance.

NOTES ON SET THEORY

The basic principles of set theory are few in number. A statement of these principles is necessary as use has been made of them in the discussion on mass spectral interpretation.

A set or class may be defined as a collection of objects called elements of the set. Specifying the elements in a set defines that set. Set theory is the study of the relations of sets to one another and to their subsets; the application of set theory provides a means of defining classes of objects in a very precise manner and of establishing those relationships which exist between various groups of objects i.e. the ions of a mass spectrum.

All sets, regardless of what other elements they may contain, have as a member the null or empty set which is designated by the symbol Ø. This is quite different from the set which contains zero as an element.

Two sets are equal when they contain exactly the same elements or members; the order in which the elements are arranged in the two sets undergoing comparison is not important. Subset may be formally defined as any set which is contained in a given set. Any set may be a subset of itself; however, the special case of proper subset is defined as a subset which contains fewer members than the parent set. The number of possible subsets in a set is given by 2^n, where n is equal to the number of elements in the parent set.

The universal set (the set containing all possible subsets or elements in a given category) is different for each specific collection of conditions which we describe.

The complement of a given set Z, designated as Z', contains all those elements in the universal set U, of which Z is a member, which are *not* indeed members of the set Z itself. The complement of the universal set U is seen to be the null set Ø and is designated as U'. Any two sets under consideration are said to be disjoint if they do not have any elements in common.

Two further categories of sets may be defined in the following simple terminology:

1) equivalent sets—sets which contain the same number of elements although not necessarily the same ones.

2) ordered set—a set in which the elements thereof are arranged in a serial relationship based on some pre-defined rule.

These latter two categories are particularly suitable to application in the treatment of mass spectral data.

References

1. Knock, B., Wright, D., Kelly, W., and Ridley, R.G., *17th Annual Conference on Mass Spectrometry and Allied Topics*, A.S.T.M.—E-14, Dallas, Texas, 18–23 May 1969.
2. Jurs, P.C., Kowalski, B.R., and Isenhour, T.L., *Anal. Chem.*, **41**, 21 (1969).
3. "Catalogue of Mass Spectral Data", Serial No. 44, Thermodynamics Research Centre Data Project, Thermodynamics Research Centre, Texas A and M, University College Station, Texas.
4. Ötvos, J.W., and Stevenson, D.P., *J. Am. Chem. Soc.*, **78**, 546 (1956).
5. Mann, J.B., *J. Chem. Phys.*, **46**, 1650 (1967).
6. Reed, R.I., *Rev. Port Quim.*, **10**, 129 (1968).
7. *ibid.*, **12**, 16 (1970).

List of participants

Lecturers

Professor Dr. Jacques E. Collin
Université de Liège
Institut de Chimie
1b Quai Roosevelt
Liège
Belgium

Dr. Francisco Mendes
Secção de Geologia
Faculdade de Ciências de Lisboa
Lisboa
Portugal

Professor Dr. M. F. Laranjeira
Lab. Calouste Gulbenkian de Espectro-
  metria de Massa and Física Molecular
Instituto Superior Técnico
Lisboa
Portugal

Prof. Dr. Maria Alzira Almoster Ferreira
Lab. Espectrometria de Massa
Instituto Superior Técnico
Lisboa – 1 – Portugal

Miss Maria Teresa Robert Lopes
Faculdade de Ciências
Lisboa – 2 – Portugal

Dr. A. G. Sharkey, Jr.
Bureau of Mines
4800 Forbes Avenue
Pittsburgh
Pennsylvania 15213
U.S.A.

Dr. Charles Merritt, Jr.
U.S. Army Natick Laboratories
Natick
Massachusetts 01760
U.S.A.

Professor Morton E. Wacks
Department of Nuclear Engineering
The University of Arizona
Tucson
Arizona 85721
U.S.A.

Dr. A. Quayle
Shell Research
Thornton Research Centre
P. O. Box No. 1
Chester
U. K.

21   Reed

Dr. George Lester
Imperial Chemical Industries Ltd.
Dyestuffs Division,
Blackley
Manchester 9. – U. K.

Dr. N.R.Daly
A. W. R. E.
Aldermaston
Berkshire – U. K.

Dr. R.I.Reed
Chemistry Department
University of Glasgow
Glasgow W.–2
Scotland – U. K.

Mr. Robert D.Graig
Vacuum Generators, Ltd.
Charlwoods Road
East Grinstead
Sussex
U. K.

Dr. Karsten Levsen
(c/o Prof. Dr. H.D.Beckey)
Institut für Physikalische Chemie
  der Universität
5300 Bonn (Rh.)
Wegelstr. 72
West Germany

Other participants

Mr. Anders O.Pedersen
Dept of Organic Chemistry
University of Aarhus
DK 8000 Aaarhus C
Denmark

Dr. Charles Larsen
Piletoften 9
Snekkersten DK 3070
Denmark

Dr. Sven-Olov Lawesson
Dept of Chemistry
University of Aarhus
DK 8000 Aarhus C
Denmark

Dr. Bernard Cabaud
6, rue Hélène Boucher
69 – Bron
France

Dr. François Guinot
53 D, rue Faubourg St. James
F – 34 – Montpellier
France

M.Jean-Michel Lavigne
20 rue Lucile
Toulouse 31
France

Dr. Josef Weigl
Bot. Inst. TH
61 Darmstadt
Germany

Mr. Theo H.Schweren
Inst. f. Phys. Chemie u. Kolloidchemie
5 Köln
Zülpicher Str. 47, Abt. Prof. Hummel
Germany

Dr. Anastassios Varvoglis
Lab. of Organic Chemistry
University of Thessaloniki
Thessaloniki
Greece

Mr. Apostolos C.Frangos
43 Ellanikou Street
Athens (501/1)
Greece

Mr. George S. Topalis
Brufa 1
Thessaloniki
Greece

Miss Farzaneh Mohammadi Tabrizi
38, Daylaman Street
Jam Shidabad Avenue
Teheran
Iran

Dr. Jehuda Yinon
Isotope Department
The Weizmann Institute of Science
Rehovoth–Israel

Dr. Rafael Shnitzer
Isotope Department
The Weizmann Institute of Science
Rehovoth–Israel

Dr. Francesco Pietra
Istituto di Chimica Generale
Università di Pisa
Via Risorgimento 35,
5600 Pisa
Italy

Mr. Lionello Pogliani
Via Arnolfo 48
50121 Firenze
Italy

Mr. Anton Sinnema
Technische Hogeschool
Lab. voor Organische Chemie
Julianalaan 136
Delft,
The Netherlands

Dr. Frans J. Gerhartl
Faculty of Wis- en Naturkunde
Dreihuizerweg 200
Nijmegen
The Netherlands

Mr. Age R. Hansen
Frydensgate 3 oppg. A III
Oslo 5
Norway

Mr. Dag Eriksen
Kurud, 1400 Ski
Norway

Mr. Dagfinn Aksnes
Chemical Institute
University of Bergen
5000 Bergen
Norway

Mr. Erik Wulvik
Kjemish Institutt
Universitetet i Bergen
Bergen
Norway

Mr. Per Albriktsen
Chemical Institute
University of Bergen
5000 Bergen
Norway

Prof. Dr. Alberto José Correia Ralha
Laboratório da Polícia Científica
Rua Gomes Freire
Lisboa
Portugal

(Miss Dr. Eng). Conceição González
Centro de Estudos de Química Nuclear
Faculdade de Engenharia
Porto
Portugal

Mr. Joaquim Gomes da Silva
Instituto Português de Oncologia
Palhavã
Lisboa
Portugal

Eng. Carlos Pulido
CEEN–Centro de Estudos de Química
  Nuclear
Instituto Superior Técnico
Lisboa–1
Portugal

Mr. Jorge da Silva Mariano
Centro de Estudos de Química Nuclear
Faculdade de Ciências
Coimbra
Portugal

Prof. Eng. Luís António Aires-Barros
Instituto Superior Técnico
Lisboa – 1
Portugal

Mr. Manuel Aníbal Ribeiro da Silva
Laboratório de Química da Faculdade de
  Ciências
Porto
Portugal

Mr. Manuel Sarmento Bravo
Caixa Postal 815 – C
Universidade de Luanda
Luanda
Angola

Mrs. Maria Áurea Isidoro da Cunha
Laboratório Calouste Gulbenkian
  de Espectrometria de Massa
Comissão de Estudos de Energia Nuclear
Instituto Superior Técnico
Lisboa – 1
Portugal

Miss Maria Eduarda Miranda Guedes
Centro de Estudos de Química Nuclear
Faculdade de Engenharia
Porto
Portugal

Miss Maria Eugénia Fronteira e Silva
Laboratório Calouste Gulbenkian
  de Espectrometria de Massa
Comissão de Estudos de Energia Nuclear
Instituto Superior Técnico
Lisboa – 1
Portugal

Mrs. Maria Fernanda Chancerelle
  Manchete
Chemistry Laboratory
Faculdade de Ciências
Lisboa
Portugal

Mrs. Maria Helena Pereira da Conceicão
Junta de Energia Nuclear
Laboratório de Física e Engenharia
  Nucleares
Sacavém
Portugal

Mrs. Maria Manuela Carvalhas
Instituto Gulbenkian de Ciência
Centro de Biologia
Oeiras
Portugal

Mrs. Maria Olinda Braga
Laboratório Nacional de Engenharia
  Civil
Lisboa
Portugal

Miss Maria Teresa Ruan Pera
Universidade de Luanda
Luanda
Angola

Mr. Raul António David Gomes
Laboratório de Técnicas Aplicadas
  à Mineralogia
Alameda D.Alfonso Henriques, 41–4º. E.
Lisboa
Portugal

Dr. José Font
Instituto de Química
Patronato de Investigación Científica y
  Tecnológica
Barcelona
Spain

Prof. Dr. José Maria Saviron
Facultad de Ciencias
Universidad de Zaragoza
Ciudad Universitaria
Zaragoza
Spain

Prof. Dr. Juan Yarza
Facultad de Ciencias
Universidad de Zaragoza
Ciudad Universitaria
Zaragoza
Spain

Miss Maria Francisca Reig Isart
Chemistry Department
University of Barcelona
Spain

Dr. Melih Erkmen
Technical University of Istanbul
Electrical Engineering Faculty
Department of Meteorology
Gumussuyu
Istanbul
Turkey

Dr. A.J.Luchte, Jr.
Bendix Scientific Instruments Division
1775 Mt. Read
Rochester, N. Y.
U.S.A.

Miss (Dr.) Joyce Wiebers
Dept. of Chemistry
Purdue University
Lafayette, Indiana, 47907
U.S.A.

Dr. Lawrence Verbit
Dept. of Chemistry
State University of New York
Binghamton, N.Y. 13901
U.S.A.

Dr. Richard Walter Rozett
Fordham University
Chemistry Dept.
New York, N.Y. 10458
U.S.A.

Dr. Theodore Axenrod
Building 10, Room 7 N 306
National Institute of Health
Bethesda, Maryland 20014
U.S.A.

Mr. Albert G.Maclean
Chemistry Department
University of Glasgow
Glasgow, W–2
Scotland
U.K.

Miss Anne H.Tennent
Chemistry Department
University of Glasgow
Glasgow, W–2
Scotland
U.K.

Mr. Bourdonais
GEC–AEI, Ltd.,
Scientific Apparatus Sales Department
Barton Dock Road
Urmston,
Manchester
U.K.

Mr. Donald H.Robertson
Chemistry Department
University of Glasgow
Glasgow, W–2
Scotland
U.K.

Mr. G.W.Ball
Finnigan Instruments
Ebberns Road
Hemel Hempstead
England
U.K.

Mr. John E. Williams
Field Tech. Ltd.
London (Heathrow) Airport
Hounslow
Middlesex
U.K.

Miss Joyce Mellor
21 Lennox Street
Edinburgh, EH 4 1PY
Scotland
U.K.

Dr. L. Fraser Monteiro
Chemistry Department
University of Glasgow
Glasgow, W–2
Scotland
U.K.

Mrs. (Dr.) M. Lourdes Fraser Monteiro
Chemistry Department
University of Glasgow
Glasgow, W–2
Scotland
U.K.

Mr. Terence L. Threefall
20 Ingestre Road
Forest Gate
London, E. 7
England
U.K.

Author Index

327

Compound Index

Subject Index

343

Acknowledgments

The editor, publishers and relevant authors gratefully acknowledge the permission given to reproduce the material below.

Fig. 1, p. 13; P. F. Knewstub and The Cambridge University Press, *Mass Spectrometry and Ion Molecule Reactions*

Fig. 3, p. 21; H. D. Hagstrum and The American Institute of Physics, *Physical Reviews,* Volume **59**, 1941, p. 4018

Fig. 3, p. 152; Don Villarejo and The American Institute of Physics, *Journal of Chemical Physics*, Volume **48**, 1968

Fig. 5, p. 154; Elsevier Publishing Company, *Journal of Mass Spectrometry and Ion Physics*, Volume **2**, 1969, p. 231

Figs. 2, 4, 6, 7, 9, and 10, and Table 3, pps. 151, 152, 159, 160, 163, 164 and 161; K. Siegbahn and Almqvist and Wiksell, *E.S.C.A., Atomic, Molecular and Solid State Structure Studied by Means of Electron Spectroscopy*

Figs. 6 and 9, p. 275 and p. 279; R. M. Teeter and W. L. Mead. Reprinted from *Analytical Chemistry*, Volume **39**, No. 14, December, 1967, p. 1834 and Volume **40**, No. 4, April, 1968, p. 745. Copyright by The American Chemical Society and reprinted by permission of the copyright owners.

Fig. 16, p. 285; G. Spiteller and Springer Verlag, *Monatshefte für Chemie*, Volume **97**, 1966, Part 4S p. 1064

Fig. 22a, p. 290; N. C. Rol and The Institute of Petroleum, *Advances in Mass Spectrometry*, Volume **4**, E. Kendrick Editor, 1968, p. 221

Figs. 24 and 25, p. 295; K. D. Schuy and Friedr. Vieweg and Sohn, *Zeitschrift für Instrumentekunde*, Volume **75**, 1967, p. 190

Fig. 26, p. 297; N. H. W. Addink and Adam Hilger Ltd., *XII Colloquium Spectroscopicum Internationale*, Exeter, 1965, p. 651

Fig. 27, p. 297; J. M. McCrea and The American Institute of Physics, *Applied Spectroscopy*, Volume **23**, 1969, p. 55

Fig. 28, p. 298; J. Whitehead and The Society for Analytical Chemistry, *Analyst*, Volume **91**, 1966, p. 418

Fig. 29, p. 298; P. F. S. Jackson and The British Ceramic Research Association, *Conference of British Ceramic Research Association*, October, 1968

Fig. 30, p. 298; R. E. Honig and The New York Academy of Sciences, *Annals of the New York Academy of Sciences*, Volume 137, Article 1, Table 3, p. 280, R. E. Honig; Reprinted by permission.